Homo Sapiens in Decline

Gerhard Kraus was born in Bohemia (Sudetenland) but came to Britain in 1938 as a refugee from Hitler and is now a British national. During the war he researched into harvesting methods and later served in the Royal Fusiliers. His interest in internationalism led in, 1950, to the organization of the Peoples World Convention held in Geneva. Some time after this he accepted an invitation to help industrialize Nigeria and established successful building, manufacturing and publicity enterprises. Six years ago he went into semi-retirement in order to complete the present book, which was originally conceived in 1941. In 1970 he co-founded, with C. E. Joel B.Sc. the anthropological quarterly, *The New Diffusionist.*

Mr Kraus is the father of five children.

Gerhard Kraus

Homo Sapiens in Decline

A Reappraisal of Natural Selection

The New Diffusionist Press

First published by the New Diffusionist Press 1973
39 West Street, Great Gransden, Sandy, Bedfordshire

ISBN 0 7121 5601 1
ISBN 0 7121 5602 X

Made and Printed in Great Britain by
Northumberland Press Ltd., Gateshead
Set in linotype Pilgrim

ACKNOWLEDGEMENT
My thanks are due to Mr Stephen Stein
for his valuable help in preparing this book.

Contents

Introduction

More than a hundred years ago Darwin wrote *The Origin of Species* and explained the mechanism of natural selection. Since then the scientific community has generally accepted the influence of natural selection on animals, but has paid little attention to its effect on the human species. This book intends to show that natural selection plays a crucial role for man, and that the recognition of this fact forces us to question many established beliefs and conclusions which hitherto seemed well founded. Among them is the view that we are the apex of all living creation and that animals are inferior. This notion of human superiority must now be abandoned. Other, even more profound ideas concerning the higher meaning of life and man's ultimate purpose on Earth and in the universe are also challenged.

In this respect natural selection, in association with genetics, allows us to observe that man has basically the same biological function as a fly or an elephant – that is, to carry a tiny germ cell, which perpetuates itself by simple cell division, into the next generation. The astonishing discovery is thereby that nature (seen metaphorically) should have taken all the trouble to evolve a sophisticated carrier called 'man', or alternatively a 'giraffe' or a 'whale', merely to perpetuate a microscopically small single cell, while a simple organism like a protozoon could have continued to do the job just as effectively. In such a biological process, exclusively devoted towards the preservation of tiny germ cells, no intelligently directed purpose towards a higher end, nor any religious motivation can be discovered. Nature's achievement, therefore, in having evolved the higher animals, including man, is like someone who has constructed an intricate computer occupying an entire building to solve the arithmetical problem of adding one and one.

Darwin was well aware of the dilemma arising from the discovery of natural selection. In 1856, he wrote, 'What a book a devil's chaplain might write on the clumsy, wasteful, blundering, low and horribly cruel works of nature.' And in 1890 he wrote, 'My theology is a simple muddle. I cannot look at the universe as the result of blind chance, yet I can see no evidence of beneficient design, or indeed of any design of any kind in details.'

By presenting us with natural selection, Darwin demonstrated that organic evolution proceeds haphazardly and without divine or any other guidance. Geneticists, by showing that hereditary changes are caused by chance mutations, have confirmed this. Accordingly, any assumption about directive evolution towards higher and better life-forms breaks down. Instead, what emerges as the mainspring of animal life is the mere maintenance and perpetuation of the germ cell. In evolutionary terms this means the preservation of species. Once we speculate beyond this biological reality of species preservation, which forms the essence of organic evolution, the supposition that there is a logical *a priori* motive in nature or the universe has to be abandoned.

It follows that the preservation of our own species should have been the aim and end of human collective endeavours at all times. Instead, man inadvertently abandoned this objective when, with the beginning of civilization, mankind became obsessed by certain religious and philosophical ideas now exposed as biologically harmful.

A proper evaluation of natural selection shows that man's acquirement of a super-brain is both the cause for his exalted position in nature as well as that of his downfall. Symbolically seen, from the moment that man began to taste the fruits of knowledge (by creating the rudiments of culture), he was already doomed. And why? Because of the particular way natural selection works.

Before the advent of genetics it was generally assumed that man was the product of progressive evolution which would eventually elevate him to an almost godlike stature. This view was based on the acceptance of Darwinian natural selection plus Lamarckian inheritance as the motivating forces of evolution. Now that we must perforce exclude Lamarckian factors as scientifically obsolete, and instead couple natural selection with genetics, it turns out that this type of progressive evolution is a myth. What we find is that a rigorous natural selection is necessary merely to keep species on existing levels of biological efficiency. Such evolutionary stagnation amongst animal species is the rule in nature, while evolutionary changes are infrequent. The latter result from the preservation of those rare gene mutations which prove beneficial to survival. Decrease in the rigour of natural selection will, on the other hand, lead to a proliferation of harmful gene mutations eventually resulting in extinction.

In the case of man the invention of fire, protective clothing, and other cultural aids diminished natural selection. Life was made more comfortable, so that individuals who would formerly have perished could now survive and propagate the species. Thereby a process of biological degeneration was engendered, which intensified as culture expanded. Its existence is evidenced by, *inter alia*, a decline of the brain size of *homo sapiens*, noticeable since Neanderthal and Cro-Magnon times about 35,000 years ago. However, due to our biological malady being compensated by scientific and technological aids, the fatality of our biological decline has remained largely undetected. Only recently has it dawned on us that our supposed super-brain seems to be unable to cope with our cultural heritage, which is now rebounding on us with a vengeance in the guise of the side effects of applied science and technology. The perilous ecological situation thereby produced is aggravated by another fruit of culture and civilization, namely our political heritage. This presents us with a world divided into many sovereign nations, rent by the cross-currents of conflicting ideological, racial, economic, and pure power interests, all militating against world political unification, and consequently also impeding global pollution control.

The problem we face is that mankind is involved in a longstanding biological decline from a formerly higher level and is simultaneously threatened by the ecological crisis. Let us assume that the ecological crisis could be resolved and that mankind would increasingly occupy itself with efforts to combat its biological recession. The only way to achieve this would be to subject mankind to a more rigorous natural selection and even this could not reverse the trend, but would merely maintain the *status quo* at the reduced biological level on which mankind finds itself at the time. Moreover, to achieve this aim, which is comparable to a man grabbing a precariously rooted tree while falling into a precipice, we would have to return to the life of the simple prehistoric food collector and hunter. It is inconceivable that modern man will be prepared to take such a culturally retrogressive step deliberately. But if he does not take it, there is nothing that can arrest his biological decline and eventual extinction. All he can try to do is merely to slow down the process.

Part One

Natural Selection: The Biological Impact

I

Natural Selection and Evolution

Most biologists now agree that one of the most potent factors in the evolution of life, and in biological activities generally, is natural selection.

Superficially observed, natural selection seems an indifferent and ruthless power. A hundred years ago this was emphasized by Herbert Spencer's dictum of 'the survival of the fittest', which showed the battle for survival as a victory for the strong and superior. The weak and helpless would be crushed. Applied to humanity it came rather close to Hitler's ideas. As will be shown later, natural selection works immeasurably more subtly. Though it frequently uses tooth and claw, it can also act gently and with restraint.

The French naturalist Lamarck was one of the first to modify the view that the various species of animals and plants had been separately created. His apparently reasonable explanation was that nature itself (or the conditions of life) moulded and shaped living beings like an invisible potter. In this process, organs in animals would also become modified by their greater use or disuse. For example, Lamarck assumed that an animal living perpetually in the dark would eventually become blind; the giraffe, by stretching out its head to reach ever higher branches of trees would thereby permanently lengthen its neck. Lamarck also postulated that any adaptive change acquired during life-time could be passed on to future generations by heredity. Thus Lamarckian evolution became synonymous with the inheritance of acquired characteristics.

How such an inheritance could be biologically accomplished remained a mystery. In any case, Lamarck's theory was still compatible with the idea of an almighty God and Creator. God could be considered as the potter, who in the guise of nature shaped all life.

It appears that Darwin initially accepted Lamarck's viewpoint in its entirety. But when he later studied the breeding of domesticated animals and plants, it occurred to him that man, by deliberate selection of better specimens, was able to modify the biological structure of the race. How the process could be applied to organisms living in a state of nature remained for a long time hidden from him. Darwin records in his diary that in October 1838 he came across Malthus's essay 'On the Principles of Population'. It at once struck him that, due to the high ratio of increase among all organic beings throughout the world, there would be a struggle for existence. 'It follows,' wrote Darwin, 'that any being, if it vary however slightly in any manner profitable to itself, under the complex and sometimes varying conditions of life, will have a better chance of surviving, and thus be naturally selected.'[1]

It was the doctrine of Malthus applied to the whole of the animal and vegetable world. But Darwin did not relinquish his Lamarckian outlook entirely. Although he came to consider natural selection as the primary influence in the evolution of life, he admitted that Lamarckian factors were also active. He even developed the hereditary theory of 'pangenesis', a pure speculation, to indicate how acquired characters might be inherited.

The Darwin–Lamarckian mixture of evolutionary theory remained more or less static until the beginning of this century. A continuing and close study of heredity, reinforced by the rediscovery of the Mendelian laws of inheritance, led gradually to the rise of the science of genetics. This involves the study of the genes and their constituents as the ultimate units of heredity. Genetic knowledge denies the possibility of the inheritance of acquired characters and excludes them as a contributory factor in evolution.

It is only in the last few decades that natural selection has gained universal recognition as the dominating factor in organic evolution. Today it is upheld by most biologists.

The term natural selection is, however, deceptive because it creates the impression that there is a deliberate process at work. But this is not the case. Darwin himself was repeatedly challenged on this. He wrote the following in his defence:

> It has been said that I speak of Natural Selection as an active power or Deity; but who objects to an author speaking of the attraction of gravity as ruling the movements of the planets? Everyone knows what is meant and what is implied by such metaphorical expressions; and they are almost necessary for brevity.
>
> In the literal sense of the word Natural Selection is a false term, but whoever objected to chemists speaking of the elective affinities of the various elements? And yet an acid cannot strictly be said to elect the base with which in preference it combines. So again it is difficult to avoid personifying the word nature; but I mean by Nature only the aggregate action and product of many natural laws, and by laws the sequence of events as ascertained by us.[2]

There is really no deep mystery about natural selection; stripped of its frills it can be reduced to the interaction of three principal factors:

1. *The high rate of increase in progeny* common to most forms of organic life (procreation).

2. *The inborn (genetically induced) capacity of living organisms to vary* from their parent forms (heredity).

3. *The struggle for existence*, between different individuals or species of animals whereby the less variable variations are eliminated. The means of elimination is natural selection. The result is the chance survival of the more viable or fitter individuals. This means that Herbert Spencer's phrase, 'the survival of the fittest', is admissible with certain reservations. They are that 'fittest' does not merely mean superiority in brute force, but includes such characteristics as intelligence, the altruistically orientated parental instincts and others.

Each of these three factors will now be separately considered.

1. The High Rate of Increase

What now appears a commonplace was by no means obvious in Darwin's time. Darwin, in *The Origin of Species*, recognized that

> ... all organic beings naturally increase at such high rates that, if not destroyed, the earth would soon be covered by the progeny of a single pair. Slow-breeding man has doubled in twenty-five years, and at this rate, in less than a thousand years, there would literally not be standing-room for his progeny.[3]

The present population explosion is a pointer in this direction. Linnaeus, as quoted by Darwin, calculated that if an annual plant produced only two seeds (and there are no plants which are less productive) and their seeds produced two the following year, and so on, in twenty years there would be a million plants.

Darwin reckoned the elephant to be the slowest breeder of all known animals.

> It generally begins breeding when thirty years old, and continues to do so till ninety years old. It bears six young in the interval. Accordingly, after a period of from 740 to 750 years, there would be nearly 19 million elephant alive, descended from the first pair.[4]

An investigation of the upper one-inch layer of an area of soil near Washington, D.C., showed that it contained more than 1 million microscopic animals and 2 million microscopic seeds per acre. Further, microscopic organisms found in the uppermost eighteen inches of sediment in a Californian estuary numbered more than 3 million per acre. Bacteria, which abound in many media (a gram of soil may contain several hundred million), were not included in the count.[5]

Sir John Russell has pointed out that in reproduction bacteria may split into two every twenty or thirty minutes. F. Johnis, a German bacteriologist, estimated that a single cell (if the increase were not halted) could in the course of forty-eight hours produce a colony 1,000 cubic metres in size, big enough to fill a hundred wagons of a goods train.[6]

Darwin's conclusion was that every being during its life produces several eggs or seeds and that most of these will suffer destruction before maturity. Otherwise, due to their geometrical increase, their numbers would soon become so vast that no country could support them. As more individuals are produced than can possibly survive, there must in every case be a struggle for existence, either one individual with another of the same species, or with the individuals of distinct species, or with the physical conditions of life.

'In nature,' says Darwin, 'there can be no artificial increase of food, and there is no prudential restraint from procreation. Although some species may increase in numbers, all cannot do so, for the world would not hold them.' It is the doctrine of Malthus applied with manifold force to the whole animal and vegetable kingdom.[7]

2. *The Inborn Capacity of Living Organisms to Vary* (*heredity*)

While Darwin's preceding deductions presented no particular complications, it remained a mystery to him and his contemporaries how hereditary variations could be produced and transmitted. It was only with the advent of genetics that the veil on this subject was gradually lifted.

If we adhere strictly to the principle of natural selection and exclude the possibility of Lamarckian inheritance in the transformation of species (which is another expression for evolution), then it must be assumed that evolution proceeds merely by chance. Consequently hereditary variations must also be assumed to result from a chance process. This condition is fulfilled by the behaviour of the genes within the chromosomes of the germ cell.

The cell is the basic unit of all living matter. Only a few generations ago it was thought of as mainly an enclosed sac of protoplasm or living jelly with a few finer structures outside the cell nucleus. Today, with the help of electron microscopy, biologists know the cell is no empty blob but a densely occupied factory containing many structures doing many jobs.

The chromosomes, which are the main ingredients of the cells nucleus, look like a string of flat, irregularly shaped discs of varying diameter and thickness. Each of these discs represents a gene, formerly considered to be the ultimate unit of heredity. About twenty-five years ago it was discovered that the genes consist of long molecules of D.N.A. (deoxyribonucleic acid) formed of links, or nucleotides. Just as a computer may do its 'thinking' in terms of a binary code (one with only two numbers) so there are only four chemical 'signals' within the D.N.A. links.

It is now known that the nucleotides link up selectively inside the genes, adenine on one chain links only with cysosine. They are paired, two to a link, in such a way that when the links are mated they form a twisted structure, a double helix. Thus when the gene goes into action, by splitting into single strands, each strand contains the blue-print for assembling its complementary strand to reproduce the original gene.

Some viruses can be assumed to consist of a single chromosome covered with protein containing 6,000 to 7,000 nucleotides.[8] The several chromosomes of higher organisms contain many hundreds of thousands of them. Latest estimates of genetic material within human cells amount to 6 billion (6,000,000,000) of such units, not all of them functioning. The University of Wisconsin has reported (June 1970) the first complete synthesis of a gene.

Generally, gene properties are as follows: the self-reproduction of a gene is a process of self-copying. Occasionally a gene is subjected to some unusual stress and undergoes a sudden permanent change. Such genes of modified type are called mutations. When genes reproduce their doubles and create replicas they accurately reproduce any new features (mutation) in their own daughter genes.

Seen in isolation a gene consists of a closely connected group of different molecules of ultra-microscopic size, while a 'gene-mutation' is principally a fortuitous chemical change within it. Gene mutations can be caused by microchemical accidents, resulting from thermal agitation, reactions of metabolism that get out of control, by high energy radiation like X-rays, or by the influence of specific foreign substances. The effect of gene mutation rarely, if ever, depends upon the type of environment or the physiological state in which an animal lives. Neither is the mutation an adaptive response to such a situation. On the other hand a mutation of a gene within a chromosome will generally also influence the other genes located in it.

Being accidental, most mutations are harmful to the organism, but the smaller their effects, the less likely they are to be harmful. Hence those very rare mutations which happen to be useful and survive and take part in the evolution of a species usually represent small steps. This confirms Darwin's assumption that evolution tends to proceed by slow gradations under the influence of natural selection. Besides mutations in individual genes, inheritable changes can also be caused by the addition, subtraction, or the rearrangement, of groups of genes.

Once it was thought that sexual propagation represented a major breakthrough in the evolution of life, and that it was a sign of higher organization. But this view had to be abandoned since it was discovered that many species of bacteria and other simple forms of life also can procreate by sexual fusion.

When considering sexual reproduction in animals it is also important to stress the dual function of the fertilized germ cell (or egg). Once fertilized, one part of the egg produces all the body cells, which remain attached to each other and form the individual. The other portion remains dormant in the germ-tract until sexual maturity is reached. When this happens the hitherto dormant germ-tract will, by single cell division, produce the new fertilizing cells, i.e. the sperm of the male or the egg of the female. This must however be preceded by what is termed reduction division.

In sexual propagation the main source of variation is the recombination of chromosomes (contributed by both parents) which embodies an enormous number of adaptive possibilities. This process has frequently been compared to dealing from a pack of cards which has been reshuffled. Additional variations of an evolutionary impact are caused by gene mutation, going beyond the possibilities of recombination. If mutations cause adaptable variations of the organism, they may form the cornerstones of new evolutionary trends. But if mutations cause non-adaptive variations the individuals produced will be unfit to survive.

All such gene and chromosomal behaviour completely accords with the Darwinian tenet of variation, which is as true now as it was more than a hundred years ago. Darwin wrote:

> Even the slightest variations from whatever cause if they are in any way profitable to the individuals of a species, will help the preservation of these individuals, and will generally survive. Because of this occurrence of the many individuals of any species which are periodically born, only a small number can survive.

Lately, some doubts have been expressed by some geneticists, including Professor Waddington, about the general validity of evolution by natural selection in the neo-Darwinian sense (i.e. Darwinism minus Lamarckism).

They contend that natural selection is not quite haphazard when bringing about evolutionary changes. Although they have not expressly said so, their views involve the assumption that there is a directive influence in evolution – something akin to divine interference. In scientific terms this might be called genetic pre-adaption, but it really involves a return to Lamarckian ideas of inheritance. In practice, however, what appears to be pre-adaption can be explained by the use an animal can make of an already inborn range of genetic possibilities.

3. *Struggle for Existence*

In nature, says Darwin, the 'struggle for existence' includes the dependence of individuals on each other (social capacities) and also the success in leaving progeny.

> Two canine animals, in a time of dearth, may be truly said to struggle with each other to decide which shall get food and live. A plant on the edge of a desert struggles for life against the drought. A plant which annually produces a thousand seeds, of which only one may reach maturity, can be truly said to struggle with the plants of the same and other kinds, which already clothe the ground.[9]

Darwin also noted that the struggle for existence will usually be most severe between the individuals of the same species 'because they frequent the same districts, require the same food, and are exposed to the same dangers'. He continued:

> Between varieties of the same species, the struggle will almost be equally severe. For example, if several varieties of wheat are sown together, and the mixed seed is re-sown, those seeds which best suit the soil or climate or are more fertile will supersede the others. They will produce more seeds, and in a few years will supplant the other varieties.[10]

Rhodes, in *The Evolution of Life,* refers to other factors which play their part in the struggle for existence:

> A severe winter may kill countless animals and plants; shifting sand-dunes will bury the vegetation of coastal areas, while new types of plants will re-populate the areas; mass deaths of wild animals may result from severe summer droughts; summer heat may bring a plague of flies; a sudden freeze may bring catastrophic death to lake fish. But one should be careful not to over-simplify the relationship between life and its environment, which is extremely intricate. It embodies the interaction of countless physical factors (temperature, pressure, light, density of water, salinity, winds, daily variations, seasonal changes, etc.) including the even more complex and subtle influences of the organisms that shape and maintain life.[11]

Among individuals of the same species, another aspect of natural selection sometimes comes into play which Darwin has called 'Sexual Selection'. It is the struggle between males for the possession of a female. The result is occasionally death to the unsuccessful competitor or, more frequently, few or no offspring.

In sexual selection the most vigorous males will generally leave most progeny, and often victory will depend not so much on vigour, but on the adversary having inferior weapons. Darwin wrote:

> A stag without horns or a cock without spurs would have little chance of leaving numerous offspring. Sexual selection, by always allowing the victorious to breed, can give indomitable courage to the lion, length of spur to the cock, and strength to the wing of the eagle. It is the same principle by which the brutal cock-fighter selects his best cocks for further breeding. Male alligators have been observed fighting and whirling round, like Indians in a war-dance for a female's possession; male salmon fighting all day long. Male stag-beetles sometimes bear wounds from the huge mandibles of other males. Certain humenopterous male insects have been described fighting for a particular female, who sits by apparently unconcerned and then retires with the conqueror.[12]

Among birds the contest is generally more peaceful. For example, it has been shown that male birds in anticipation of mating use their singing capacities to define and defend a territory and thereby shy off possible rivals. As a rule it is not the aesthetic aspect of the male bird's beautiful plumage that attracts the female, but the mere fact that it adds to the conspicuousness of its owner.

We are attempting to set out an explanation of natural selection and evolution in as simple terms as possible, for all too often specialists have obscured Darwin's ideas behind a smokescreen of verbosity. Indeed, we rely as far as possible on many of Darwin's own time-honoured examples to illustrate our thesis, as in what follows dealing with the giraffe and the elephant. This may also be a corrective to those biologists who have written rather disparagingly about some of Darwin's early ideas.

Darwin's explanation of the giraffe's long neck is quite different from Lamarck's. Darwin observed:

> The individuals which were the highest browsers and were able during a bad season to reach even one inch or two above the others into the branches of trees would often have been preserved; but also their increased bulk would have acted as a protection against almost all beasts of prey excepting the lion. Against all these animals the giraffe's tall neck, and the taller the better, would serve as a watch tower. Hunters confirm that no animal is more difficult to approach than the giraffe. It also uses its long neck for offence or defence, by violently swinging its head armed with stump-like horns.[13]

Darwin pointed out that the elephant is rarely destroyed by predators and even the tiger in India rarely dares to attack a young elephant which is protected by its dam. Before man appeared in India or Africa some other cause must have prevented the continuing increase of the existing elephant. It is believed that insects which incessantly harassed and weakened the Indian elephant were responsible for checking its increase and the same has been assumed of the African elephant in Ethiopia. In many cases, for example in parts of South America, insects and blood-sucking bats control the numbers of the large naturalized quadrupeds.[14]

Furthermore, Darwin reminds us that the increase in number of all creatures is constantly being checked by many hardly noticeable hostile agencies; and they alone may be sufficient to cause rarity, and finally extinction. Surprise has often been expressed that such great monsters as the Mastodon and the more ancient dinosaurs became extinct. But mere bodily strength is not decisive in the battle of life. Mere size can even have the opposite effect, for in times of dearth big animals may face quick extermination because their larger food requirements cannot be met.[15]

Another factor playing an important part in the struggle for existence is mimicry – the capacity of animals to camouflage themselves against enemies by merging with the environment. Insects often resemble such objects as green or decayed leaves, dead twigs, bits of lichen, flowers, pines, excrement of birds, or insects of a different species. This resemblance is often wonderfully close, and is not confined to colour, but extends to forms, and even to behaviour. The caterpillar which projects itself motionless like a dead twig from the bushes on which it feeds is a good example.[16]

It has been observed that in some districts of South America, where a certain butterfly, *Ithomis*, abounds in swarms, another butterfly, *Leptalis*, is often found mingled in the same flock. The Leptalis resembles the Ithomis in every shade and stripe of colour and even in the shape of its wings. The resemblance is so great that even experts are continually deceived by their likeness. But when the two types are caught and closely compared they are found to be different in structure and belong not only to distinct genera, but often to distinct families.[17]

The Ithomis, which abounds in great numbers, generally escapes destruction because it is distasteful to birds and other insect-devouring animals. But this is not so with the Leptalis which imitates them externally. Thus when a variety of Leptalis resembles the common butterfly, inhabiting the same district, this variety, from its resemblance to a flourishing and little persecuted kind, has a better chance of escaping destruction from predatory birds and insects, and is consequently more often preserved and can propagate its kind.

Examples of how natural selection can create rapid adaptations of organisms to new living conditions are drug- and pesticide-resistant strains of microbes and insects.

Until the early 1930s a high proportion of deaths and much painful and ugly suffering was due to germs causing pneumonia, peritonitis, syphilis, typhoid fever, and other diseases. Such germs entered the body, settled in some tissue, fed upon it, multiplied and destroyed it. Many people recovered from such infections but the struggle was often long and painful, and recovery sometimes left the patients with permanent disabilities.

Then Domack, in Germany, discovered the action of the sulphonamides, thereby initiating one of the greatest advances ever made in chemotherapy (the treatment of disease by chemical agents). The sulphonamides when given in certain doses could kill a large variety of germs inside the body while leaving the patient wholly or virtually unharmed.

Many previously serious or fatal illnesses could now be cured rapidly before any serious damage was caused to the patient. But sulphonamides also helped surgeons and obstetricians. Childbirth became safer. Bolder operations could be performed. Some years later penicillin, the first of the antibiotics, was developed and clinically used. Other antibiotics which followed proved capable of dealing with germs that had remained immune to the sulphonamides. It began to look as if the doctors armed with anti-bacterial drugs had finally gained control over the problem of infection.

But then came the reaction. Strains of old, dangerous bacteria reappeared and refused to submit to the new drugs. In some way they had acquired an immunity. They began to give trouble, particularly in hospitals, where antibiotics and sulphonamides were so widely used that tiny concentrations of these agents were permanently present in the very air of the wards and corridors.

Bacteriologists then found that amongst any particular colony of bacteria there would always be some varieties with a slight quirk in their genetic make-up which enabled them to survive the attack of the new antibacterial drugs better than their poorly adapted relatives. If such colonies of bacteria were exposed to low concentrations of the new drugs, the weaker germs would be eliminated by natural selection and die but the tougher ones would survive. The latter then multiplied rapidly, and whole groups of new drug-resistant germs evolved having been favoured by natural selection.

In conclusion, it can be said that our investigation shows how the interaction of a profusion of offspring, hereditary variation and natural selection can lead to the emergence of new species of plants and animals. It also shows that quickly multiplying species, like drug-resistant germs, have much greater possibilities of adaptation to changing living conditions than the slower breeding animals which, like the dinosaurs and mammoths, can be overtaken by events.

A DEFINITION OF EVOLUTION

Evolution can be defined as the continuous diversification of living forms (or species) under the influence of natural selection. In this process small unicellular organisms have become transformed into such different forms of life as fish, birds, bumble bees, elephants, and *homo sapiens*. But this does not necessarily imply that there has been a qualitatively progressive development which could be considered as an advance from lower to higher and better forms. There are many relatively simple living organisms which surpass more highly organized ones in resilience and survival capacities. For example, the lowly spring-tail (*Colembola*) has existed with little change for the past 300 million years. It has outlived countless so-called higher animals such as the mighty Mastodon as well as the larger-brained Neanderthal man.

But evolution includes progressive as well as retrogressive changes. The progressive trend in evolution develops from simple to more complicated organisms and changes from older to newer forms. Examples of the latter are protozoa becoming metazoa, fish becoming amphibia, the primitive primate-brain developing into that of *homo sapiens*, or the fins of fish eventually transforming into human hands and feet. Retrogressive evolution of organs can be observed in the one-toed horse which gradually emerged from a five-toed ancestor, and also in birds like penguins which have lost the capacity for flight, or in amphibia which, having lost their legs, became reptiles. Again, though the reptiles were the first group to establish themselves fully on the land, some of their number returned again to the seas. The same return to the water had already occurred with the earlier amphibia.

In its true essence, therefore, organic evolution is a continuous process of diversification of living forms. The primordial germ-cell formed various structures around itself by gradual steps, so ensuring its own survival and continuance in competition with rival germ-cells. This competition is initiated by the capacity of the genes within the chromosome of the germ cells to mutate and attach varying structures to themselves. In their entirety these structures represent the prototypes of many different animals and they act in each case as the carriers, protectors and providers of the germ-cells from which they have arisen.

In such a process natural selection eliminates those individuals or varieties which are less adaptive. The survivors, if forming homogeneous groups, can be arbitrarily classified as species. In sexually reproducing animals, species are groups of animals which can interbreed. Different species either do not interbreed at all, or if they do so, are likely to produce only sterile offspring (for example, the sterile mule, resulting from the cross-breed of horse and ass).

Guided by these and similar considerations, Darwin called variations 'incipient species', pointing out that all living species have arisen from pre-existing species by a process of 'descent by modification', a formula which is synonymous with 'evolution'.

THE BASIC EVOLUTIONARY TREND

It can be observed that throughout animal evolution the original primordial germ-cells, due to genetic accident and by natural selection have formed, or attached to themselves, various sac-like structures. Generally the digestive tract forms the centre around which all other organs are grouped. Therefore the simplest and most generalized description of a multicellular animal is perhaps that it is a sac attached to a germ-cell. Its function is to capture and absorb nourishment so as to keep itself alive and allow the germ-cell to reproduce itself: or, expressed in a more sophisticated way, to assist in the survival of the species it belongs to.

Some of the simplest forms of life are the protozoa (although they are by no means simple in function). Taking an amoeba we find it to be a protozoic form, where the cytoplasm (i.e. the cell-body) has assumed an irregular shape with various lobes, extending from the centre like plump feet. These lobes, appropriately called pseudopodia (false feet), expand and contract at will, by this means effecting locomotion. If an amoeba comes into contact with a food particle, the lobes engulf it like the tentacles of an octopus (of which this amoeba is perhaps the primitive unicellular forerunner) and absorb it.

Another kind of protozoa often described is the 'paramecium'. Its protoplasm is about one thousandth of an inch long and shaped like a slipper. The body moves by the waving of hairlike 'cilia'. The paramecium contains an oral groove into which a current of water carrying food particles is swirled by something like an undulating tongue. Some protozoa, like 'Englena Agilos', possess a funnel-shaped opening (a gullet) leading into a wider sac-like tube which acts as the digestive tract.

The original multicellular forms of animals (metazoa) most likely arose from the incapacity of certain unicellular forms to separate after cell-division. Another possibility is that single independent cells (as in some protozoa) united to form metazoa. It is possible that none of these multicellular structures was able to survive in this new united form unless and until it incidentally assumed the shape of a sac into which water with food particles could flow. In retaining and absorbing (digesting) the food the primordial stomach was formed.

One of the simpler forms of multicellular animals is the 'hydra' (freshwater polyps). These are classified as 'Coelenterates' (meaning 'hollow inside'). They consist of a cylindrical sac, equipped with an opening mouth, which is surrounded by tentacles capable of grasping food, and effecting locomotion, representing a more advanced version of the amoeba's pseudopodia.

As evolution progressed the sac developed side branches rather like internal irrigation channels to assist the water carrying the food particles in penetrating a wider area of the organism. From this system of channels eventually emerged our blood circulation. In this respect sea water was indeed the forerunner of our blood.

When one reaches the kingdom of the worms, one can describe a worm as a tube within a tube (or a sac within a sac). The tube begins with the mouth, continues into a digestive sac and ends in the anal opening. Reptiles are of similar form, and their description as giant worms is quite fitting.

The human body possesses a mouth, continuing into a digestive sac-like tract (the stomach) and ends in the anal opening. Around this digestive apparatus our other organs and the extremities are grouped. It may be claimed that this is an over-simple explanation of the human body and that, for example, the heart or the lungs or the brain are more vital for our survival than the stomach (indeed a man can perhaps live for some time without a stomach and can be fed intravenously). But the fact remains that the stomach is perhaps the most ancient organ. It is the pivot around which all other organs were grouped throughout the whole evolutionary process, from hydra to *homo sapiens*.

SOME POINTERS TOWARDS HUMAN EVOLUTION

According to Romer,[18] the transformation of a gill arch into jaws is a natural evolutionary development, representing perhaps the simplest possible solution of developing a pair of vertebrate jaws.

Fingernails are said to be among the few primate anatomical specialities. Another useful adaptation is the position of the eyes, which, placed on the front of the head, have promoted in man the capacity for stereoscopic vision.

Another development towards man-like species, according to Howells,[19] is seen in the growing independence of the hands. Relieved from supporting the body while walking, they can be used for many other different purposes. They can supplement one another; one hand, for example, can hold a nail while the other hand swings the hammer. Howells believes that when primate hands became adapted so as to grasp movable objects, like fruits, where formerly they were used to cling to boughs and other supports, a major step in the direction of human evolution was accomplished.

Howells has further noted that in mammals who live on the ground the powers of perception are mainly concentrated in the snout and depend on smell. In animals who live in trees the sense of smell loses much of its value. The air directly above the ground is a dense world of smell, but higher up the smells dissipate. An arboreal creature gliding swiftly along a bough cannot sniff and nuzzle everything that excites its curiosity without losing its balance and falling out of the tree. But using its hands an animal may tug and pluck buds, fruits, insects, or eggs, turn them over, pull them apart, and extend the formerly limited focus of the eye-snout combination.

It may be said that organic evolution, beginning with the most simple forms of life and expanding into more complicated forms, is an unfolding of the genetic properties already inherent in primordial germ cells.

There are, for example, considerable similarities in plant and animal metabolism. This has led H. Kalmus to remark that many substances and biochemical processes are common to fungi and to man and are probably universal to all life. Many of these biochemical pathways and enzymatic reactions have been first investigated by using mutants of fungi.

Romer has argued that the problems of plants are very similar to those of animals. They can be summed up as the need to trap energy in order to perform biochemical syntheses. The real difference between plants and animals is rather in the source of their glucose. Plants are able to utilize the energy of light for a reversal of the processes of glucose oxidation (photosynthesis). In this way they can provide themselves, and also the heterotrophic world (dependence of animals on plant carbons), with the only ultimate source of organic compounds which are available in the world outside a chemist's test tubes.[20]

Romer points out that

> . . . man is probably justified in considering himself more advanced than anaerobic bacteria because he can oxidize his food all the way to carbon dioxide and water, a process that requires a more elaborate metabolism than is necessary in micro-organisms. Also very primitive organisms contain in their cells neither a nucleus nor mitochondria, whilst more advanced animals have evolved them as a contribution to the stability and efficiency of the cell.
>
> Most of us would hesitate to compare ourselves with fish, typhoid bacteria, cancer cells, or even oak trees, yet the fact is that we have very much more in common with them than we might have guessed.

Why? The forms of life that we know today have evolved over several thousands of millions of years, in the course of which they have branched out into such a host of diverse directions that their outward resemblances are remote. Yet, during the whole of this evolution, their biochemical forms have remained remarkably constant.[21]

2

The Decrease of Natural Selection and its General Effect on the Human Species

So far the decrease of the intensity of natural selection in the human species has not yet been sufficiently appreciated by the scientific community. The reason lies in the strongly held view that man is still progressing in his evolution. Many theories in anthropology and history, and some political ideologies (including Marxism) are based on this faulty assumption. Therefore the realization that there may exist a long-standing evolutionary stagnation and even degeneration in the human species is bound to put an entirely new complexion on the study of early man's history and also throw an illuminating light on contemporary events.

By taking a closer look at early human evolution it can be seen that, compared with our immediate animal ancestors, present man's bodily efficiency has in many ways declined. Evidently, at the beginning of the stone age (Palaeolithic) bodily deficiencies were compensated by the evolution of greater brain capacities. But at some stage along this evolutionary path, brain evolution apparently also came to a stop.

This is by no means a new discovery. Several writers, amongst them Darwin, have indicated that such a possibility exists. Gordon Childe pointed out that

> ... already at his first appearance in the Pleistocene up to now man has been inadequately adapted for survival ... in any particular environment. His bodily equipment is inferior to that of most animals to cope with prevailing conditions. He has not, and probably never had, a furry coat like the polar bear's for keeping in the body's heat under cold conditions. His body is not particularly well adapted for escape from the exceptionally fleet of foot, and would be left behind in a race with a hare or an ostrich. He has no protective colouring like the tiger or the snow leopard, nor bodily armour like the tortoise or the crab. He has no wings to offer escape and give him an advantage in spying out and then pouncing upon prey. He lacks the beak and talons of the hawk and its keenness of vision. For catching his prey and defending himself, his muscular strength, teeth and nails are incomparably inferior to those of the tiger.
>
> This poor bodily endowment is compensated by a large and complex brain. Thereby man has been able to design protections against

climate and weather. He has produced arms and weapons of attack and defence. As they can be adapted and adjusted they are superior to furs, or teeth, or claws.[1]

WHY ANIMAL SPECIES REMAIN CONSTANT

Practically all species of living organisms except man and his domestic animals are involved in an intensive struggle for existence. But, in spite of this, most species remain as a rule biologically constant over long periods of time. It may therefore be suggested that evolutionary stagnation (which is the preservation of the constancy of a species) is the rule in evolution.

In nature, due to the great potential increase of progeny to which all forms of organic life are subject, every available living space is most of the time already occupied by various species. Therefore 'selection pressure' is, as a rule, always high. And even if a species should come to occupy a new environment where there is little competition from other species, its own numbers will soon increase to such a point that the ensuing struggle for existence between the individuals of the same species will quickly restore a population balance.

There is now fairly unanimous agreement amongst geneticists that in the majority of cases gene mutations are harmful to life. The reason is that all the genes of an individual must act in harmony to produce a well-balanced organism. In each generation this harmony is preserved through the weeding out of the less adapted individuals. The adversely affected individuals, as we have shown, are the result of fortuitously produced gene mutations. Thus any inherited irregularity of the genes will generally have a harmful imbalancing effect on the individual concerned. The rare exceptions when gene mutations cause new and useful adaptations are the building stones of evolution.

The natural constancy of wild species cannot be due to the absence of natural selection. Natural selection is as effective in maintaining the qualitative stability of a population as it is in changing it. S. Maynard Smith[2] points out that the typical or average members of a population are usually the best adapted and the extremes in any direction are at a disadvantage. Darwin cites the example of certain winged insects living near the sea-shore. Those with long wings are usually carried far out into the sea by strong winds and perish; those whose wings are too short will have difficulty in escaping from enemies. The insects with wing lengths intermediate between the two extremes will have the best chance of survival.

According to Rendel, duck eggs which failed to hatch had almost the same average size but were far more variable than eggs which did hatch. (The same principle has been demonstrated by Kearn and Penrose for the birthweight and survival of human babies.)[3] Darwin observed that species will as a rule not only remain constant over long periods of time but they will also frequently remain so under changed conditions of life. 'Certain species,' he says, 'have migrated over vast spaces and have not become greatly or at all modified.'[4]

There are living species and even genera of animals for which evolution seems virtually to have ceased for hundreds of millions of years. One existing genus of *Lingula*, according to Rhodes, has a history going back 500 million years. How can such a group, let alone a genus, survive so long unchanged? The clue to the survival of this and similar examples, like the sponges and jellyfish, seems to lie in their simplicity.[5]

Other groups which have persisted over long periods of time are the lungfishes; the modern crocodiles, which differ little from crocodiles of the Jurassic period; and the opposums of South America, which closely resemble the small pouched mammals that had a world-wide distribution millions of years ago. Another is a lizard-like reptile (Spendou) now found only in New Zealand, but closely resembling an animal of Triassic age recently discovered in Gloucestershire by L. Robinson.[6]

Soil insects like spring-tails undergo practically no metamorphosis; they are wingless, and seem to be survivors from the days before insects had developed wings. An insect closely resembling the present-day spring-tail has been found in the lower Devonian rocks in Scotland. It dates back some 300 million years, i.e., some 40 or 50 million years before the arrival of the earliest winged insects.[7]

Turtles are an example of an unchanged reptile. The turtles, once within the shelter of their armour, became the conservatives of the reptilian world,' says S. H. Romer.

> The ruling reptiles grew to dominate the reptilian scene, but the turtles persisted unchanged. The dinosaurs passed away, and the mammals took their place, but the turtles went calmly on their placid way. Now man dominates the scene, but the turtles are still with us. And if, in the far distant future, man in turn disappears from the Earth, very likely there will still be found the turtles plodding stolidly on down the corridors of time.[8]

As a rule therefore gene mutations in species living in a state of nature will all the time produce bodily variations which give their offspring differing survival chances. The majority of the varying offspring are bound to be less viable, only a few will survive, and they will be those which vary least from their parent form. Their naturally selected survivors will become the parents of the next generation, having themselves been the end result of a selective process of high intensity. Hence the parent-like survivors represent the highest qualititive standard a species can reach under prevailing circumstances.

Animals deviating from this norm will be less adaptive. Those offspring who most resemble their parents will survive, while those deviating too much from the parent form will be eliminated. In this way the constancy of species will prevail. The retention of the status quo of a species is therefore about the best intense selection pressure can produce. In other words, a species having attained a high equilibrium of biological efficiency towards an existing environment will tend to preserve it.

The general principle that emerges from such reasoning is that an intense natural selection pressure is necessary to keep animal species on a high level of biological efficiency; or, in other words, in a state of 'evolutionary stagnation'. The contrary process, resulting from an overall reduction in the intensity of natural selection, must lead to a decline in the qualitative standard of a species (degeneration) and end eventually in extinction.

HUMAN EVOLUTIONARY STANDSTILL AND DEGENERATION

According to Gordon Childe, for about 25,000 years man's bodily evolution has been virtually at a standstill. Men of the Aurignacian and Magdalenian cultures, compared with present-day man, show only negligible physiological differences, but their cultural difference is enormous. Progress in culture, in the human species, has replaced further organic evolution.[9] Ralph Solecky, describing a series of excavations in the Shanidar Valley of Northern Iraq, draws attention to a Mousterian sequence of Neanderthal remains embracing a period of about 60,000 years (2,000 generations). He found that throughout this period the Neanderthalers remained culturally stagnant. Moreover, the classical Neanderthaler skull form in Shanidar I and Shanidar II remained unchanged over a period of 15,000 years, in spite of the fact that during this period considerable climatic changes took place. Thus we have a good example of human evolutionary stagnation during a period when natural selection must have been considerably more intense than in modern times.[10]

Other students of man's history have noted only slight evolutionary changes during the last 70,000 years,[11] and Weston La Barre expresses the view that during the last 500,000 years man has stopped evolving physically in any massive or significant way.[12]

'Degenerate' can be defined as 'having departed from the high qualities of race and kind, to fall from a nobler state, to be or to grow worse' and 'degeneration' as the act or process of becoming degenerate.

In nature, as opposed to man-made conditions, true biological degeneration of species does not exist. It can only affect isolated individuals during their lifetime and natural selection will sooner or later wipe out any degenerative trait. What may therefore often mistakenly appear as the degeneration of a species should rather be called retrogression (certain reptiles having lost their legs, or birds having lost their capacity for flight). Under natural conditions retrogressing limbs are compensated by other biologically advantageous properties.

The whales and dolphins probably arose from some form of carnivorous creadont ancestor. But the hind-limbs and hair which were of little use in their new existence were lost. (In size, as well as in form, the whales are the largest animals ever to have lived. The blue whale can measure up to 120 feet long and weigh 150 tons.)

In the course of human evolution the diminution of the sense of smell located in the olfactory region of the brain was compensated by the enlargement of our visual powers (visual cortex). Many other bodily deficiencies have also been counter-balanced by a larger and more complex brain.

Man, then, has the distinction of being affected by degeneration, without organic compensation, and yet surviving, and this applies also to domesticated animals under the care of man. The compensation is supplied by cultural aids. In nature degeneration without organic compensation, and without cultural aids to take its place, must generally lead to extinction.

In man, from the time that he began to use fire, tools, and clothes to protect himself from the inclemencies of nature, emergent biological deficiencies have been compensated by these very cultural means. At the same time the intensity of natural selection was blunted in its effect. Today defective eyesight can be improved by glasses, appendicitis can be operated upon, difficult births can be assisted by Caesarean section, diabetes can be treated with insulin, etc. Thus, in general, biologically defective individuals, who in former ages would have perished without leaving offspring, are now able to survive and can continue to propagate the species.

In the evolutionary sense the incidence of biological degeneration applies of course only to hereditary defects or to defects caused by a hereditary predisposition. One can therefore define human biological 'degeneration' as a decline from a formerly higher state of physical and mental efficiency.

Nowadays human evolutionary regression in purely physical terms is not often disputed. On the other hand, there prevails a widespread belief that human brain evolution, in the form of increased intelligence, is still continuing. This belief is based on the supposition that man's increased intellect mirrors itself in his cultural, scientific and technological progress. In other words, it is maintained that cultural progress is synonymous with an increase of intellectual power. This assessment is incorrect. On the contrary, a study of the comparative intensity of natural selection, as it affects man and animals, shows that cultural progress leads to biological regression. The evidence of human fossils, relating to the size of the cranium, supports this view.

INTENSITY OF NATURAL SELECTION IN NATURE AND IN MAN

According to J. M. Smith the simplest genetic change in a population is the replacement at a locus of one allele by another. Natural selection leads to the extinction of individuals carrying the less favourable allele.[13]

Haldane calculated the number of deaths, spread over many generations, required before one allele has replaced another. He found that, unless selection was very intense, the number of deaths equalled about ten or twenty times the number of individuals in the population at any one time. Occasionally there were a hundred times this number. He suggested that the mean value is about thirty times the population number.[14]

Smith says:

> The number of selective deaths which can occur in a single generation is partly dependent on the maximum number of offspring which a pair can produce. Even if a single female can lay millions of eggs, as can a cod or herring, the vast majority of deaths will be fortuitous, not selective. Survival will be more due to luck, not because of genetic advantages. Even under intense selection, hardly more than half the deaths will be selective.[15]

Haldane maintains an intensity of selection of 0·01 is a more probable figure. If so, it would take 30÷0·01=3,000 generations of selection to replace one allele by another. Species probably differ by alleles at about 1,000 loci (that is, the number is probably greater than 100 and less than 10,000). If so, the evolution of a new species would take about 300,000 generations.[16]

It has been shown that some species have lived for millions of generations without showing any significant evolutionary change in spite of an intense selection pressure. Therefore the preceding figures indicating possible evolutionary changes must be considered as a minimum of generations necessary to effect an evolutionary change, rather than being the norm.

A comparative study of death-rates supplies the clue to the relative intensity of natural selection in animals and man. For example, according to H. Kalmus, only two out of 1,000 of some insect species will on average reach maturity.[17] This suffices to keep the species constant in numbers and stable in its biological quality. To demonstrate the intensity of selection in nature Darwin conducted the following experiment. He dug and cleared a piece of ground three feet long and two feet wide where there could be no choking from other plants. Then he marked all the seedlings of common weeds as they came up, and found that out of 375 no less than 295 were destroyed.[18]

He also pointed out that climate with its periodical seasons of extreme cold or drought seems to be the most effective of all checks determining the average number within a species. He estimated that the winter of 1854-5 destroyed four fifths (80 per cent) of the birds in his own ground. He considered this a tremendous destruction, pointing out that 10 per cent is an extraordinarily severe mortality rate in human epidemics.[19]

However, as nature overwhelmingly demonstrates, high death-rates are not indicative of evolutionary changes. In the majority of cases they keep species of animals merely constant in numbers and leave them biologically unchanged.

The intensity of natural selection in the human species or rather its lack can be deduced from an increase of life expectancies and composite death-rates as well as from increases in populations. Thereby, greater life expectancy and rises in population numbers are both indicative of diminished death-rates resulting from reduced selection pressure.

It seems then that all through history there has generally been little or no change in man's natural fecundity, and that human procreation left to its natural urge proceeds just as unrestrained as in animals. It was once believed that early man's fecundity was lower than that of civilized man, and that this was the main cause for the small size of palaeolithic (old stone-age) societies. Today this theory is no longer held. There are of course no statistics to prove the case, but indirect evidence supports the view that the palaeolithic populations had very high mortality rates. Cippola writes: 'Since the species survived we must admit that primitive man also had a very high fecundity rate. At the same time, life expectancy appears to have been very low.'[20]

Valis, quoted by Cippola, analysed fossil remains of 187 Europeans of the Neanderthal group. He found that a third died before the age of twenty, and the rest between the ages of forty and fifty. From thirty-eight Sinanthropus fossils Weideneich assessed a probable average age at death of twenty-two. Fifteen died when less than fourteen years old, three died between the ages of fifteen and twenty-nine, three between forty and fifty, and only one seems to have survived beyond fifty.[21]

From the examination of mummy cases of Roman Egypt it has been deduced that upper-class Egyptians two thousand years ago had a life expectancy of twenty-five to thirty years; in the city of Rome it was perhaps twenty, though in some Roman provinces it was probably about thirty-five. The last figure can be compared with some from the Middle Ages. For instance, from Breslau, Germany, it has been reported that the estimated life expectancy was about 33·5 years in the year of 1690.[22]

A tabulation of average life expectancies from various ages given by A. J. Harrison[23] is as follows:

	Years
Early Bronze Age	18
50 B.C.	22
Middle Ages	33
1687-91	33·5
1789	35·5
1838-54	40·9
1900-1902	49·2
1946	66·7

Cippola points out that since the Industrial Revolution the high, recurrent death peaks have tended to disappear. Death rates fell and increasing birth-rates led to an explosive growth of population. This had been due to improved sanitation and the steady progress in medical science, coupled with improved nutrition.[24]

Estimates of world population during the stone age range widely at between 2 and 20 million. Cippola suggests a figure of 5 to 10 million as being more realistic. Statistics of world population increases from more recent times (based on Cippola's conservative estimates) are as follows[25]:

A.D.	*million*	
1650	470	
1750	728	an increase of 55 per cent in 100 years
1800	906	
1850	1,171	,, 62 ,,
1900	1,608	
1950	2,377	,, 103 ,,
1955	2,528	
2000	3,727*	,, 138 ,,

(*Latest estimate 6 to 8 billion).

In all societies, the largest proportion of deaths is usually due to infant mortality. In Europe before this century and in some underdeveloped countries even now, of 1,000 newborn children, 200 to 500 generally die within a year. Many of the remaining ones die before the age of seven.[26]

Due to preventive medicine, death-rates amongst infants have been drastically reduced in most parts of the world. A. Barnett states that in England and Wales at the beginning of this century, sixty-five out of every 100,000 children under fifteen died of diphtheria. Between the wars, with some children immunized, the figure fell to about twenty-nine. In 1947, due to extensive immunization, the death-rate was down to two per 100,000.[27]

Historically, there has been a continuous decrease of death-rates in all age groups (running parallel with higher life expectancies and population increases) ever since *homo sapiens* passed from the hunter and food-gathering stage of culture to one of agriculture and civilization. This process is tantamount to a progressive reduction in the intensity of natural selection.

There is, of course, no way to assess death-rates amongst peoples of the old stone age (Palaeolithic). But once fire and clothing came into use, people of less robust health and of more delicate constitution were able to survive and to procreate children.

It seems that as life became more comfortable there was an accelerated tendency towards biological recession. This can be seen from a comparison of the mode of life under stone-age conditions and that of modern times. Under the former all energies were concentrated on eking out a living and assuring protection from ferocious animals. This was accomplished by very primitive means. It necessitated an alertness and intelligence certainly in excess of that required under civilization, where a more sophisticated arsenal of tools, weapons, implements, clothes and shelter makes circumstances less hazardous. Consequently, we must assume that during the old stone age a more intense selection pressure prevailed. Under conditions of civilization life became less and less severe. At the same time, the store of increasing knowledge, passed on from generation to generation takes the place of further brain expansion.

In literate societies, with an elaborate division of labour, the 'average' level of intelligence in terms of pure survival value is probably lower than in cultures where the problems of survival were much more immediate. The premium placed on human intelligence in the face of prolonged scarcity in the Australian desert, or at the edge of the polar ice cap, where the penalty for stupidity is death, is almost certainly greater than it is for even the most down-trodden inhabitants of Western Europe or Northern America.

One major factor which has worked throughout human history, at least for the last 5,000 years, and which can be assumed to have caused adverse biological selection towards degeneration, is conventional warfare. In primitive wars, the fittest and most courageous individuals were bound to perish in great numbers, while the majority, non-warring, weaker individuals remote from the line of combat, survived and were left to propagate the species. On the other hand the privations of the civil population have some selective effect in the opposite direction.

In organized warfare, and particularly in the case of large standing armies, a selective draft board will reject all those who fall below a certain physical and mental standard. As a consequence, the unfit ones will survive in greater numbers than those who serve on the fighting fronts. Hitler was an irony of history in this respect. His supposed objective, and that of his adherents, was the biological improvement of the Germanic races. War and soldiering were proclaimed as the most desirable professions and the most severe biological selection was applied in the case of special S.S. troops (who were chosen to father a future super-race). During the Second World War, these biologically selected troops were usually sent into action wherever the German Army found itself in difficulties. On many occasions and in conformity with Hitler's strategy such S.S. troops were forbidden to retreat. Consequently they were killed in their thousands. In this way Hitler achieved the opposite of what he had set out to accomplish by eliminating those who had been biologically selected and the preserving of those who were considered as biologically inferior.

In modern civilization many biological deficiences that natural selection would most likely have eliminated in former ages survive by medical intervention. For example, many babies nowadays are born by what is known as Caesarian operation which, by abdominal incision, serves to enlarge the uterus so that the baby can be delivered. A too narrow uterus must therefore be considered as a congenital malformation. In former times, when sterile operational practice was still unknown, any pregnant woman thus handicapped would have died. Today Caesarian operations help the survival of many babies who are likely to pass this deficiency to mothers of future generations. Acute appendicitis, or rather the inheritable tendency towards it, presents a similar case.

Gerald Leach has pointed out that roughly speaking, one in every forty babies is stillborn or dies within a year because of congenital defects or birth injuries. Worse still, one in every twenty-five babies is born with a physical or mental handicap and lives. Not all these survivors are gravely handicapped, and on present reckoning only a quarter of them, or one baby in every hundred, is known to be the victim of a purely genetic defect.

He adds:

> These figures are bound to grow as scientists uncover the mostly mysterious causes of the remaining three quarters [of defective births], some of which are certain to turn out to be genetic. In fact, it is widely estimated that for every known genetic defect there are another three or four to be discovered.[28]

Many deficiencies which may be due to genetic factors, and others where there is only a hereditary predisposition, are known. They include diabetes, many eye defects, colour-blindness, dental caries, dental malocclusion, haemophilia, mongolism, androplasia (dwarfishness), gout, Huntington's chorea, fibrositic disease of the pancreas, phenylketuronia, gastric and duodenal ulcers, rheumatic fever, pyloric stenosis, certain types of hip-dislocation, deafness, and many others.

Thanks to medical science, many of these genetic disorders are now being successfully treated. As a result the proportion of individuals within a population with hereditary defects is bound to increase, and these defects will be passed on to future generations.

At first sight, this does not seem to matter much. In modern society, in spite of a lower level of biological efficiency, people live generally healthier lives than in former ages. This is partly due to better hygiene, sanitation, improved nutrition, and, above all, to the influence of modern science and technology. The question is only how far this biological deterioration can be left to develop before the damage caused to the populations concerned becomes irreparable.

3
The Problem of Biological Regeneration

There can then be little doubt that mankind is involved in a process of biological degeneration. The question is whether this creeping degeneration can be arrested at all, and, if so, how? This brings to mind past attempts to improve the heredity of man which emanated from Francis Galton (a cousin of Charles Darwin). It was during the middle of the last century that Galton started the eugenic movement in England. Eugenics meant the application of the methods of animal breeding to the human species. Early eugenics was inspired by the vision of breeding a physically and intellectually superior type of human specimen. This principle led eventually to the outrage of Hitlerite racism with the endeavour to breed a Nordic master-race.

Significantly enough, the early eugenists completely ignored the influence of natural selection on the human species. Their main point was that breeding among the upper classes, the aristocracy and professional people in particular (who statistically produced relatively few children) should be encouraged. The lower classes, on the other hand, should be restrained from having offspring. Thus it was assumed that economic and social status was concomitant with better heredity. The early eugenists also disregarded the fact that the quest for money, status and fame was not necessarily a biologically desirable trait of human nature.

In Galton's time the labouring classes and the poor lacked proper nutrition and sanitation, and newborn babies were exposed to the most severe health hazards. The wealthy, on the other hand, were able to live healthier and more comfortable lives, and provide better educational facilities for their children and give them a better economic start in life. As a consequence the children of the élite, even if hereditarily inferior in various respects, would survive and produce children, while defective offspring among the underprivileged had little chance to survive and breed. Consequently, the poorer strata of the population, having been subjected to a much more rigorous natural selection, must through the ages have formed a gene pool that was superior in quality to that of the wealthy classes and the aristocracy.

This does not mean that the backward populations and the economically weaker classes could have served as a gene pool from which to produce a superman. But increased natural selection tended to keep this human majority less susceptible to degeneration. Even where inheritance does matter, as it does to a large extent with intelligence, stature or 'beauty', individuals rarely breed 'true'. These qualities are controlled by many genes, and in a genetic shuffle of sexual reproduction they often crop up in surprising new combinations. To achieve any worthwhile 'improvements' the group would have to go in for very intensive, selective breeding for a long time.

The geneticist Kenneth Mather has estimated that if a group kept strictly apart from the rest of the population, bred only 'bright' with 'bright', and threw all the dimmer children out of the clan, then they would probably double their average intelligence quotient. But it would take them twelve generations or about 300 years to do so.

Another geneticist, John Maynard Smith, has considered a more plausible scheme. Suppose, he says, that a group of people isolated themselves and that most of the women had one child by a genius. After three generations the mean intelligence quotient of the group would have risen by fifteen points. If the founder women had an average intelligence quotient of 100 then most of their grandchildren could get into a university. Maynard Smith rather acidly remarks, 'This hardly seems to justify the establishment of another new cult.'[1]

Summing up these possibilities, Gerald Leach writes:

According to present-day genetic theory, if as many as one in ten of all women have half their children by male donors with intelligence quotients fifteen points higher than their own, in a generation the average intelligence quotient of the whole population would rise by about half a point, which is not reliably measurable by an intelligence quotient test. The intelligence quotient cultists would also run into the difficulty that one can improve only one or two characteristics at a time and almost invariably at the expense of others. Somewhere ahead of them would lie a fate similar to that of the modern manufactured bacon pig; layered with nice lean meat, but stupid, docile and unable to walk more than a few yards.[2]

Other examples of how the breeding of special traits usually affects animals adversely are given by F. A. Beach, who writes:

Breeding of domestic chickens for high egg production has produced behavioural deficiencies of various kinds. Although hens of some lines are excellent layers, they have almost totally lost the normal tendency to brood the eggs once they have been laid. The maternal behaviour of sows of different inbred lines of swine is strikingly different. Females of one line are so aggressively protective of their young that they cannot be approached during the lactation period. Sows of a second genetical line possess such weak maternal interest that they frequently kill their litters by stepping or lying on the young.

From this and similar considerations it becomes obvious that the breeding of a perfect animal or of a superman is an illusion. Considering that human stock, through thousands of years, has become riddled with inheritable defects, due to a decrease of natural selection, it is even doubtful whether human stock can be improved at all by artificial means. The reason is that man-made selection can hardly be presumed to achieve the subtlety and intricacy of the natural process.

Darwin recognized this more than a century ago and genetic discovery has not altered the position:

> It may metaphorically be said that natural selection is daily and hourly scrutinizing, throughout the world, the slightest variations; rejecting those that are bad, preserving and adding up all that are good; silently and insensibly working, whenever and wherever opportunity offers, at the improvement of each organic being in relation to its organic and inorganic conditions of life. We see nothing of these slow changes in progress, until the hand of time has remarked the lapse of ages, and then so imperfect is our view into long past geological ages that we see only that the forms of life are now different from what they formerly were.[3]

Perhaps if mankind would return to the harsh and humble conditions of life that prevailed during the thousands of years of the old stone age, it would be possible to recapture to some degree the force of natural selection and thus regenerate human stock biologically. But to do so, mankind would have to give up most of its cultural gains and return to the social life and an economy of small family groupings.

In nature, as opposed to man-made conditions, the need for eugenic measures does not arise; natural selection, working on its own resources, eliminates all biological defects solely on the principle of species preservation. Species survival is in itself already the highest biological criterion by which an animal species can be judged under natural conditions. This immediately gives a pointer to human biological measures. Man must aim at the maintenance of the present existing biological (evolutionary) level and its protection against a further encroachment of degenerative hereditary influences.

And here it must be asked whether greater intelligence – this so much treasured property of modern man – is in itself of such great and decisive survival value for mankind in its present state as is claimed. Our biological history, compared with that of most of the animals in a state of nature, indicates the contrary if seen in terms of pure survival value. There is no criterion by which to show that exceptional intelligence, as exemplified by high performance in intelligence tests, is biologically beneficial or even desirable in the long run. On the contrary, it appears that social and humane impulses, which stem from our more instinctive urges, may have more survival value.

The objective we must have in mind is that of the loving father (or mother) whose main concern is to care for the family, who has compassion not only for his children but for his fellow men and for animals as well and who lacks the rather morbid urge to make the accumulation of wealth and the quest for fame and status his life's purpose. If we can produce and maintain this kind of human specimen as the average world citizen of today and tomorrow, the prolonged existence of mankind can probably be assured.

What then are the chances of introducing biological measures? At the present time, birth control methods for restraining population growth (including the extensive use of the contraceptive pill, inter-uterine devices, abortion and sterilization) are being encouraged in many parts of the world. All these measures are carried out haphazardly and without any consideration of selective factors. The question is how selectivity could be introduced into measures of birth control.

The most effective method designed so far – one which could be used as a starting point – is genetic screening. This involves the examination of genetic material, taken from the living cells of individuals, for potential inheritable defects.

A London geneticist who had performed the test and aborted the foetus of a mother with a high risk of a mongol baby said recently: 'The time will come when it will be routine for all older mothers. These are the ones most likely to have a mongol child.'[4] Gradually the technique will spread to innumerable other chromosomal deformities like Turner's syndrome, in which dwarfing, webbing of the neck, and absence of ovaries are common.

To see the possibilities of selective birth control in their true perspective we must redirect our attention to the population explosion. It has been estimated that by the end of this century the world's population will reach 7,000 million (7 billion). This huge mass of people is bound to strain the Earth's natural resources to an unprecedented extent. To slow down this population explosion and keep it within reasonable bounds, measures of world-wide birth control seem to be unavoidable. This raises the possibility of combining measures of birth control with the prevention and elimination of hereditary defectiveness, thus making birth control selective.

But this worthy objective is not easily realized. Firstly, we cannot asume that selective measures can keep step with present population increases, even if the most energetic action is taken. Furthermore, it happens that the economically poor countries of the world show increases in their populations at an immeasurably higher rate than the economically rich ones. It has been estimated that of every 100 people born between 1965 and 1985, eighty-five will live in the poorer countries.

This means that the greater part of the world will soon become deeply involved in the overwhelming problems of food supply, urban proliferation and unemployment. Under such conditions measures of selective birth control are sure to be pushed entirely into the background. Regrettable as it may seem, some selective effect will be achieved by the then inevitable Malthusian apocalypse of famine, civil strife, and war.

4
The Specific Effect of Natural Selection on Human Intelligence

It is of considerable interest for our argument to find out in what way the decrease of natural selection has affected the human brain and how this, in turn, has affected human society. We shall begin by comparing the sizes or brain capacities* of animals and man, and then consider to what extent such cranial capacities are an indication of the brain's functional value.

Here are some comparative sizes of animals as given by W. de La-Barre in *The Human Animal*:

Stegosaurus	70 *grams*
Sheep	130
Tiger	290
Elephant	5,000

Brain weights of apes are as follows:

Gibbon	80–140 *grams*
Orang-Utang	300–390
Chimpanzee	330–400
Gorilla	330–600

Comparative average human brain sizes are:

Modern Man	1,410 grams (*Encyclopaedia Britannica*, 1961 edn.)
Neanderthal and Cro-Magnon Man	1,650 grams (H. F. Osborn, *Men of the Old Stone Age*)

What can be deduced from the cranial capacity or the weight of the brain? According to von Bonin, a leading authority on brain evolution, the weight of the brain is a very poor indicator of its functional value. He found that there was no apparent correlation between the average cranial capacity and the cultural status of a race. Among brain-types variations between say 1,200 c.c. and 1,800 c.c. can be found without correlation with any known psy-

* Brain capacities are given in grams of brain weight or in cubic centimetres. In converting grams (g.) into cubic centimetres (c.c.), an addition of about 4 per cent is needed.

chological character. However, a certain minimum of brain is necessary for the normal functioning of an individual. Keith[1] believed that there was a cerebral Rubicon, which he put at 800 c.c.; only when the cranial capacity is greater than this are we justified in assuming an intelligence and that ability to learn which appears necessary for human status. Dart[2] later tried to show that the strict adherence to this value, calling anything below 800 c.c. infrahuman and anything above it manlike, was not justified. He quoted a number of examples of human beings with brains that were far below this arbitrary standard, notably, examples of Bantus with capacities of 511 c.c., 519 c.c. and 561 c.c. who were able to function in a human way as herd boys and farmhands.

Pearson[3] showed that some very gifted persons, including Leon Gambetta, Anatole France, and Franz Joseph Gall, had small brains of about 1,100 grams, while Byron and Dr Johnson had brain capacities of about 2,000 grams. He concluded that correlation between brain sizes and mental capacity was not significant.

Whether this argument can be extended to an evolutionary series is another matter. We know far too little about the bodily proportions of fossil forms. Clearly the brain stands in a certain relation to the rest of the body, and the rest of the body is still largely hidden from us.[4]

That there is no definite correlation between brain size and brain quality (i.e. intelligence) is emphasized by von Bonin. He says that the inside of the brain case gives at most only a vague clue about brain quality. The brain is sheathed in the *dura mater* and is seen as through a thick veil, so that very little can be discerned. He believes that anthropologists who have given a detailed analysis of the fissuration of the frontal lobes of fossil-men are probably wrong. However, it has been shown that the fissures of the human brain are quite variable and these variations do not appear to throw any light on the mental characteristics of the bearer. 'The analysis of fissures of fossil brains as to brain qualities is largely wasted effort', is von Bonin's conclusion.[5]

What about other features of the brain? Certain scientists have based human brain superiority on the particular development of the frontal lobes. This view is rejected by von Bonin who believes that the importance of the frontal lobes of the brain has often been vastly exaggerated. A careful study by Chapman and Wolff (1959) supports this view. They say that it is the amount of tissue lost, rather than where it is lost, that determines whether the higher functions will be disturbed or not.[6] This is so with monkeys; and the human brain does not differ in this respect from the brains of lower forms.

The more reliable brain indices show that the frontal lobes of fossil forms differ little in size from those of modern man. The frontal lobes of the fossil forms are relatively as large as those of modern man and larger than those of the macaque (a type of Rhesus monkey). The fossils of the Upper Palaeolithic (later part of the old stone age) present structures similar to modern man, except that their brains are generally larger. But in both types of man the indices for the various parts of the brain do not differ greatly.

It must be emphasized that these views about human brain evolution, although largely accepted on the authority of von Bonin, are nowadays held by the majority of brain anatomists. It appears that specific human brain qualities can neither be deduced from differences in brain volume nor from the shape of the brain itself. It is here that an evaluation of the varying intensities of natural selection, as expressed in differences of death-rates, can serve as a guide.

First of all, natural selection assures the retention of the existing level of biological quality. Secondly, to effect evolutionary changes intense selection must be coupled with the preservation of gene mutations which have selective value. As greater intelligence was obviously of selective value in the early evolution of the human species, and as in former times selection pressure must be assumed to have been more intense than it is nowadays, brain evolution in the qualitative sense must have taken place in man.

Such an improvement in human brain power suggested by fossil remains correlates with a definite increase in the average size of the cranium, at least on the evidence of Australopithecus, Peking Man, and perhaps Swanscombe Man (Woman) and other sources. Thus, although Neanderthalers and Cro-Magnons reached the greatest average brain size (as a group) ever recorded, between about 70,000 and 35,000 years ago, the peak in brain quality was perhaps already reached before that time. Contemporary man cannot, therefore, claim better brain qualities than some of his stone-age predecessors.

All this may seem rather hypothetical. But the fact remains that as human culture advances, natural selection diminishes in intensity. It must be assumed that a further brain improvement in an evolutionary qualitative sense, at least since Cro-Magnon times, seems to have been in the highest degree unlikely. Fossil evidence supports this view as will be shown in the following pages. The least we can assume, disregarding for the moment the possibility of human degeneration, is that human brain capacities, when linked to natural intelligence, have remained stagnant for several tens of thousands of years. In other words, our human ancestors of thousands of years ago already possessed the same brain capacities and had reached a similar level of intelligence to that of modern man.

THE EMERGENCE OF HOMO SAPIENS

The earliest beginnings of the human species, as far as the establishment of a continuous line of descent is concerned, are still wrapped in mystery. The reason is that the fossil record of human remains is very incomplete and well-preserved specimens are rare. Nevertheless, the position is more favourable today than it was only several decades ago.

As methods of chronological determination become more reliable a more definite picture of continuity is emerging where formerly there was only a hazy outline. While we must still expect some radical adjustments in fossil interpretation, there is now a certain unanimity amongst palaeontologists about the general path which human evolution has taken.

At the time of writing, the claim for the most ancient man-like fossils goes to a specimen called Kenyapithecus-Africanus, discovered by Dr Leakey, who estimates its age at 20 million years. He claims that the specimen was man-like, although other scientists believe it was more of an ape.

The next in age is Rhampithecus discovered by two American scholars, E. L. Simmons and Pilbeam. Its estimated age is between 8 and 15 million years, and it is believed to be a direct predecessor of Australopithecus (Southern Ape). There exists also a jaw fragment discovered by Patterson, with an estimated age of 5 million years. F. C. Howell found some fossils in the Omo River Basin, which he estimated to be 4 million years old.

All these discoveries are from areas in southern Africa and eastern Africa south of the Sudan. Some jaws and teeth of human fossils from southern Europe and from mainland China, whose age is put at 6 to 8 million years, have also been found. At the time of writing no detailed reports have been published about them.

The Australopithecines are the next group of man-like species on the time scale. They are a widely dispersed group of South African origin. Whether they include such types as Zinjanthropus and Paranthropus is still in dispute. While there is no unanimity amongst anthropologists about such an inclusive grouping, we accept it for the time being in the absence of a more plausible explanation.

Zinjanthropus was discovered by Dr Leakey at Olduvia Gorge,

Tanzania. He describes him as very closely related to Australopithecus, probably a sub-genus. Some individuals of this species have been dated at 1·82 million years. Another fossil find that interests us here is *homo habilis*, a contemporary of Zinjanthropus and a supposed member of the genus *homo*. He seems to have resembled our own species more closely than other fossils of similar age. There are still differences of opinion as to whether Australopithecus was a vegetable- or a meat-eater, whether he used tools or not, and whether he knew how to use fire.

An interesting point is that Australopithecus, or its variations, has a wide geographical distribution and that some of the finds indicate links with Java Man (Sinanthropus). Similar finds have been made in Lastolil, in East Africa, and in Sangiran (Megonthropus), in Java. They show that all these fossils belong to the genus Australopithecus.

Pithecanthropus (Peking Man) and Sinanthropus (Java Man)

Both these early forms indicate distinctive stages of human evolution (i.e. *homo erectus*). With their appearance definite evidence of the use of fire can be associated with human fossils for the first time.

According to W. Howells, there is little difference between the skeletons of Peking Man, Java Man, and our own. 'Peking Man had rather thick walls to his thighbones, but he showed none of the distinctions in shape seen in the Australopithecines.'[7]

It has also been found that Peking Man's brain was larger than Java Man's by about 200 cubic centimetres and possibly more. The measurable skulls found averaged about 1,075 c.c., with a suggested figure for males of about 1,150 c.c. This compares with a modern male average of 1,450 c.c.

Rhodes describes Peking Man as still ape-like in appearance. 'It was a rather squat creature, about five feet high, with heavy beetling brows, a retreating forehead, and powerful jaws. He walked erect, and the nature and arrangement of his teeth are man-like.' Peking Man's age is estimated at about 500,000 years. Java Man is considered to be of lesser antiquity with an age of about 350,000 years.

Howell reports that an important find of a lower jaw was made in 1963 by Chinese scientists at Chenchiao in the Lantian District, Shensi Province, and in 1964, about fifteen miles away, they discovered some face parts, a tooth, and a skull-cap all from the same skull. It is believed that both are of earlier date than Peking Man, and correspond with the oldest Java specimens. Their cranial capacity is only about 750 c.c., close to Koenigwald's estimate for the Djetis Skull (Java).

Howells, regarding this as a most satisfying discovery, says:

> We see a contemporary and obviously close relative of Java Man well up in the Chinese mainland; it is fairly clear that only one general kind of man occupied the whole Far East during the Middle Pleistocene. We now have not merely the uncle of Peking Man, in the earliest of the Java Man, but apparently his father as well, in the new Lantian Men. We see a small degree of progress between the earlier and the later specimen.

Neanderthal Man (*New Man*)

In 1856 in a cave near Dusseldorf (Germany) a fossil skull-top and the bones of a human specimen were discovered. It was named 'Neander' (the Greek for new man) and accordingly the place was later named 'Neanderthal'. Adherents of the evolutionary theory believed initially that Neanderthal Man represented the intermediate link between ape and *homo sapiens*.

This theory has since been abandoned. One theory is that the Neanderthal is a parallel species to *homo sapiens*; another that some *homo sapiens* intermingled with Neanderthalers and that amongst our ancestors there are many hybrids of the two. The specifically Neanderthal type of man is supposed to have died out about 35,000 years ago.

Neanderthal remains have since been discovered in western Europe in Belgium, France, Germany, Italy and Spain and later also in North Africa, Palestine and Russia and recently in Rhodesia and the Cape of Good Hope. It is believed that the later skulls belonged to different 'races' and are the result of geographical isolation.

Some of the Neanderthal remains are difficult to distinguish from modern man. In Palestine a skeletal population from Mt Carmel presents individuals that combine varying degrees of Neanderthal with modern man's traits. Such a population can be interpreted as a hybrid population or as one which is ancestral to both. Their cranial capacity is variously measured between 1,400 c.c. and 1,800 c.c. with an average approximately 100 c.c. above that of modern man.

Neanderthal–Australian connections are indicated by Professor Elliot Smith. He refers to a number of primitive structural features amongst the Australian aborigines which suggest affinities with the species *Neanderthalensis*.[8]

It is believed that the Tabun Neanderthal Woman is about 40,000 years old. 'In other words', says W. Howells, 'we have, in fairly late times, a Neanderthal population (Tabus, Amud) giving way to homo sapiens.'[9]

Modern Man

Traces of what can be termed 'Modern Man' have been discovered in various parts of the world and some of the remains are of considerable antiquity. Among the oldest are the finds at Swanscombe and Steinheim (both women), which have been estimated by Howells to be as much as 400,000 years old. The Kamjara skulls, supposed to have been contemporaries of the early Neanderthalers, are estimated to be 60,000 to 70,000 years old.

A specimen of Cro-Magnons in the Grotto des Enfants is described by Broca to be 6′ 4½″ high (193 to 194 cm.), the face very broad in relation to its height, the skull with an unusually large brain capacity. Vernon estimates that the five large male skulls of Cro-Magnon type at Grimaldi have an average capacity of 1,800 c.c.

The fossil remains of the late part of the old stone age show that the inhabitants were to all intents modern men. Among them von Bonin lists the skeletons from Les Eyzies, described by Broca in 1868, the skeletons from the Grotto de Grimaldi, near Monaco, the skull from Chancelade, the skeletons from Oberkassel near Bonn, the youth from Aurignac, and the skulls from Predmost and Krapina. In von Bonin's opinion there are few differences between the Upper Palaeolithic and most modern races. For example, only the *foramen magnum* is both longer and broader than it is in modern man. The skull capacity is about 40 to 50 c.c. greater than in almost all modern races. These are about the only differences that even very careful biometric research can unearth.[10]

The Swanscombe and Steinheim Skulls

These skulls were found at Swanscombe in England and at Steinheim in Germany. They are of considerable importance for our argument.

Morant, in the official report containing a detailed comparison of these two skulls, concludes that they are considered to be of the modern type and clearly distinguishable from the Neanderthalers. Rhodes writes:

> Two fragments from the gravels of the Thames at Swanscombe in Kent are associated with Acheulian implements. Although these bones are relatively thick they nevertheless show a very strong similarity to those of modern man. They indicate that *homo sapiens* may have

existed for at least 200,000 years. The Steinheim skull near Stuttgart, may be of comparable age. It has a smaller brain and prominent brow ridges which are reminiscent of Pithecanthropus.[11]

The Swanscombe and Steinheim skulls are the only human fossils of any consequence from the early part of the long span between the end of the second glacial period to the beginning of the fourth. Their age is estimated by W. Howells at 400,000 years.[12] According to Oakley's dating the Swanscombe skull is only 250,000 years old.[13] The Steinheim woman's brain size is estimated at 1,150 c.c. 'There is', says Howells

> an impression of modernity quite different from either the Neanderthals or such thick-skulled creatures as the man of Peking. At one time the Steinheim skull was supposed to be of much later date than Swanscombe and more closely related to the Neanderthals. After a generation of speculation and doubts it was concluded that the Swanscombe and Steinheim skulls (both women) are much alike and that they are contemporaries.

He continues:

> Human development during the last half million years seems to have arrived at key stages; final enlargement of the brain in the Neanderthals, and a lightening up of the structure and face in *homo sapiens*. But here we have brain cases of nearly modern size and shape in the Middle Pleistocene. And that, as long as her face escapes us, is the main meaning of the Swanscombe Woman.[14]

Describing two other important fragments from the Fontechevade Cave in Central France (Charentre), Howells says that the brain was probably over the 1,450 c.c. mark, that is, above the average of European men of today (the sex of the owner cannot be determined).

> ... In its outlines the vault of this skull is indistinguishable from a modern man, although it is definitely thick and is particularly broad across the back just where it is broken. In other fragments the brow ridge is clear to see: it is slight and just about what would be typical of a woman of today.

It has been frequently asserted that the apparent decline in pure brain size has been compensated by some increase in brain quality, though how such an improvement could have been accomplished while the human species was going through a period of decreasing natural selection pressure has never been indicated. For example, S. H. Osborne, remarks:

> Generally, increase in brain-power is correlated with increase in brain volume. Some extinct human groups (as evidenced by fossilized remains) have had a larger brain than present-day man. Some apologists have

denigrated such facts without valid counter evidence by maintaining that the relatively smaller brain of contemporary man is of better quality, i.e. to make up for the loss of volume.[15]

The same problem has been considered by L. L. Brace and M. F. Ashley Montague from the point of view of natural selection. They say that up to a point there was a selective advantage in favour of those individuals who were best able to master the necessary cultural adaptations.

> In considering that there is a rough correlation between brain size and learning ability, it is obvious why cranial capacity increased during the greater part of the Pleistocene. Eventually a level was reached where the efficiency of the cultural adaptive mechanism was such that it benefited even those individuals who were not completely able to master it. At such a point, individuals who could manage with the necessary minimum requirements of culture would have just as good a prospect for survival as those who could master their culture with ease ...
>
> One might even speculate that if any differences in intelligence are discovered in the future they will be in inverse proportion to the efficiency of the cultural adaptive mechanisms of the group concerned. In highly effective cultures even the dully-witted can survive and reproduce, provided they master at least the rudiments of their language and the mechanics of the social system. This requires a mental minimum and although modern technology has succeeded in lowering it somewhat, there is a limit below which effective participation in one's culture is so much reduced that the chances of such a person to marry and leave offspring (for genetic perpetuation) are minimal.[16]

They add that increase in intelligence apparently had some adaptive value up to the Neanderthal stage. Since then a balance seems to have been maintained with, if anything, a slight reduction of the premium placed on brains and possibly a slight reduction in the average intelligence, particularly of people who live under more advanced cultural systems. Unless, therefore, man is able to manipulate his genes no increase in intelligence can be predicted.

THE INFLUENCE OF DEATH-RATES

Earlier it has been shown that an intensive natural selection pressure is necessary to keep species on a healthy level of biological efficiency. The intensity of the selection pressure can be judged from the percentual elimination rates (death-rates before maturity is reached) to which species in nature are normally subject.

For example, elimination rates (death-rates) among some insect species allow only two out of 100 to reach maturity (i.e. a death-rate of 98 per cent).

Darwin observed elimination rates in seedlings with a survival of twenty out of 100 (death-rate 80 per cent). Haldane and Zeuner have reckoned with general survival rates of 0·1, a survival of one out of 10 (death-rate 90 per cent) excluding deaths due to accidents.[17]

Reliable figures of human death-rates can only be taken from more recent historical times. From former ages the only guidance comes from estimated life expectancies and population increases. Darwin assumed that death-rates during epidemics rose to about ten in 100 (10 per cent). C. M. Cippola quotes child mortality rates amongst primitive agricultural communities of between 50 and 75 per cent. He gives the following figures for recent times: France in the year 1800—19 per cent, in 1950—5·3 per cent; India in 1900—20 per cent, in 1950—13·7 per cent; U.S.A. in 1900—16·2 per cent and in 1950—3·3 per cent. Comparative death-rates for all age groups are: France between 1801 and 1805—2·63 per cent, between 1851 and 1855—2·41 per cent, between 1905 and 1909—1·95 per cent and in 1950 1·27 per cent; India between 1905 and 1909—4·3 per cent and in 1950—1·61 per cent; U.S.A. between 1905 and 1909—1·54 per cent and in 1950—0·96 per cent; China in 1950—2·5 per cent.[18]

According to the same writer, life expectancies of food-gathering peoples during the stone age have been deduced from the fossil evidence of specimens from eighteen to twenty-two years old.[19] This indicates a death-rate of much below 60 per cent. But if a death-rate of 80 to 90 per cent amongst animals under conditions of nature generally achieves no more than the maintenance of a constant level of biological efficiency (i.e. stagnating), anything below 50 per cent is unlikely to lead to biological improvement.

Recent human evolution cannot, therefore, have been one of biological progress in a qualitative sense. Furthermore, a decrease of natural selection has led to the spread of genetically inheritable deficiencies. This, plus the adverse selection resulting from wars, is a sure pointer towards human degeneration.

In spite of such signs of biological regression, mankind appears to continue to survive quite comfortably. Everywhere there is a general rise in health standards, people grow taller and live longer, populations increase. In the field of sport, records show that there is apparently no barrier to improved performance. In science and technology, the intellectual expansion appears almost unlimited.

Yet there is no evidence to link these improvements with an improved heredity of the human race. The causes of cultural advance are based rather on the better use of an already existing inborn range of intellectual potentialities, aided by scientific and technological advances.

It appears safe to suggest that human evolutionary history has taken approximately the following course. Firstly, we note the emergence of a man-like species from some ape-like ancestor between 1 and 3 million years ago. Then occurred the evolution of human types, with need for longer childhood care requiring a longer protective period: this was perhaps the prime necessity for greater intelligence leading to a consequent brain expansion (based on genetic adaptation). There followed further brain expansion leading to the use of primitive tools and hunting weapons, made of wood, bones, and crude unworked stone. Such implements took the place of greater hardiness and led to the survival of types already physically regressing but increasing in brain power. Then, from as early as 500,000 years ago, men began to make use of fire and protective clothing. The consequence was a further drop in the intensity of natural selection making continuing brain evolution unecessary and rather unlikely.

During this period the gradual decrease in biological quality (robustness) was made good by increased cultural (technical) adaptation. These intellectual achievements were attained in spite of unchanged brain-power and were chiefly based on accumulated knowledge. About 8,000 B.C. there are signs indicating the beginnings of agriculture and, in the fourth millennium B.C., of urban literate civilization.

As a consequence of the subsequent cultural progress under civilization natural selection pressure decreased still more, being perhaps accompanied by a slight decline in the general human intellectual level.

During modern times (between the eighteenth and twentieth centuries) there occurred an explosion of knowledge spreading from Europe and becoming a world-wide phenomenon, and generating a vast scientific and technical revolution. In the writer's view this scientific and technical revolution has hidden, and is hiding, the prevailing physical, and to some degree, mental, decline of man. A vast collective stock of knowledge has now become available in which most members of society are able to participate, irrespective of their particular genetic characteristics. In such an environment there is no selective process which prevents biologically lesser-endowed individuals from surviving, marrying, and producing children. Consequently, the intellectual level of modern man must inevitably decline. Fundamentally, as L. Brace and A. Montague have pointed out, this should be relatively unimportant because of the social limitations which total mental incompetence imposes on its bearer, in so far as men and women below a certain mental level will hardly have the opportunity to marry and produce offspring.

Part Two

Natural Selection and Culture

5
Biological Evolution and Cultural History

Few people today believe that the Earth has the shape of a pancake. Yet in cultural theory the 'flat Earth' mentality still dominates, in the form of the belief in the spontaneous and independent origin of cultures and civilizations. More than a century ago Darwin, in *The Origin of Species*, destroyed 'flat Earth' thinking in biology which had led to the belief in the special creation of each organic species. In its place he put the comparative Galilean concept of the interrelationship of species and their descent from a common source. Darwin succeeded in proving that the interrelationship between species held good even where no direct evidence of physical contact could be produced. This treatise suggests that most of the cultures of this world in general and the various urban literate civilizations in particular are also interrelated and that the thesis of their independent and spontaneous generation has become untenable.

But in comparing the two processes, one relating to biology, and the other to anthropology, a word of caution is necessary. Because, in biology, the origin and interrelationship of species is determined by different factors to those which apply to the origin and interrelationship of cultural elements. In the former process the activating factors are heredity (i.e. genetic mutation) and natural selection, while in the latter it is the non-biological factor of historical accident, human contact (diffusion) and continuity which operate. And continuity again includes the growth, diversification, and even the decline of already existing cultural elements.

It can be shown that from the time that man used fire and clothing and other cultural aids, natural selection in the human species was reduced to such an extent that further organic evolution came to an end. In other words, Man's most intensive period of cultural creativeness runs parallel with a period of human evolutionary standstill and recession.

Hence, all the cultural advances made during the history of *homo sapiens* up to the present have emanated from human societies whose members have remained stagnant in a biological evolutionary sense.

BRAIN SIZE AND CULTURE

How does the principle of evolutionary stagnation affect human culture? In man, up to a certain stage, evolutionary changes were produced which culminated in *homo sapiens*. As in all other animals (except domesticated ones) these changes resulted from certain gene mutations promoted by natural selection and led to a marked expansion of the brain. It can be shown that this brain expansion was not paralleled by an increase of physical efficiency. On the contrary, as brain-power increased, man's other physical attributes were reduced in their efficiency. This was summed up by Gordon Childe when he wrote ...

> ... already at his first appearance in the Pleistocene up to now man has been inadequately adapted for survival in any particular environment. His bodily equipment is inferior to that of most animals to cope with prevailing conditions. This poor bodily endowment is compensated by a large and complex brain. Thereby man has been able to design protections against climate and weather. He has produced arms and weapons of attack and defence. As they can be adapted and adjusted they are superior to furs, or teeth, or claws.[1]

If we submit man's brain evolution to closer scrutiny we find that it failed to make further significant progress after Swanscombe times (between 400,000 to 200,000 years ago). Certainly, no further brain expansion has taken place since the height of Neanderthal and Cro-Magnon times (about 75,000 to 35,000 years ago). On the contrary, since then there has been some brain recession. This has occurred without the recovery of a lost anatomical hardiness which was reduced when brain expansion began.

Approximate average sizes of fossil man down to our time are listed in table I below.

Table I.

---

1. 2½ TO 1 MILLION YEARS AGO:

Australopithecines (Southern Men including such types as Zinjanthropus and Paranthropus).

650 c.c. according to Dart quoted by von Bonin (p. 24) (3)[6]

675–680 c.c. according to Howells[2]

750 and 800 c.c. referring to two female Paranthropus. Crassidens children, quoted by von Bonin (p. 26).

2. 500,000 TO 200,000 YEARS AGO:

Sinanthropus (Java Man)[3]	950 c.c.	(Howells, p. 261)
Pithecanthropus (Peking Man)[4]	1,150 c.c.	(„ p. 261)
„ „ „	1,300 c.c.	(Roddes, p. 253)[5]
Homo Leakeyi (Olduvai, E. Africa)	1,000 c.c.	(Howells, p. 213)
Steinheim female (Germany)	1,150 c.c.	(Male estimate 1,280, by Howells, p. 256)
Swanscombe female (England)	1,325 c.c.	(Male estimate 1,475, by von Bonin, p. 40[6])

N.B. Both 'Java Man' and 'Peking Man' are now grouped jointly under *homo erectus.*

3. 70,000 YEARS TO 35,000 YEARS AGO:

Neanderthal Men:		
Neanderthal (Child of 9 years)	1,490 c.c.	(Howells, p. 237)
La Ferrassie (France)	1,640 c.c.	(„ p. 242)
Amud Man (Israel)	1,800 c.c.	(„ p. 242)
Cro-Magnon Man:		
Grotto des Enfants (France)	1,800 c.c.	(Brocca and Vernon)
Grimaldi (Monaco)	1,800 c.c.	(quoted by Osborne, p. 292[7])

4. CONTEMPORARY MAN:

Overall average: male	1,450 c.c.	(quoted by Howells
British average: female	1,315 c.c.	(*Encyclopaedia Britannica,* 1966 edn.)
British average: male	1,456 c.c.	(*Encyclopaedia Britannica,* 1966 edn.)

Contemporary Australian aboriginals show extremes of 850 c.c. for some females and well over 1,500 c.c. for some males.

The picture that emerges from these figures shows a rise in brain capacities from about 650 c.c. in the Australopithecines (compared with 600 c.c. in adult gorillas) to 1,325 c.c. in the fairly modern type of Swanscombe female (about 400,000 years ago). From 50,000 to 35,000 years ago we find human brain capacities averaging 1,600 to 1,650 c.c. in the Neanderthalers and Cro-Magnons. From 35,000 years ago we note a gradual reduction to a world average of 1,450 c.c.

An interesting comparison arises if we contrast the merely quantitative factors of the Swanscombe (British) female of 400,000 years ago with the present average British female of 1,315 c.c. By adding 150 c.c. for the measurement of the Swanscombe female we get the proportionate brain capacity of the Swanscombe male as 1,475 c.c. These figures are thus higher than the present British averages. This points to the possibility that nearly half a million years ago there lived human types whose brain capacities were not only equal to but possibly larger than those of contemporary modern men. Assuming that there is no qualitative factor involved, we may conclude that men and women living almost half a million years ago possessed at least the same intellectual capacity as ourselves.

What the preceding table tells us in cultural terms is:

(a) from the beginnings of the old stone age, a million years ago, to about 300,000 years ago, brain expansion increased from the 650 c.c. of the early Australopithecines to Java and Peking Man's 950 c.c. to 1,300 c.c. During this interval of about 700,000 years, cultural progress advanced from a purely pebble culture to one of crudely shaped chopping tools (pre-Chellan to Chellan-Acheulian).*

(b) From about 300,000 to 70,000 years ago, roughly a quarter of a million years, brain expansion proceeded from Peking Man's 1,300 c.c. to Neanderthal's 1,650 c.c. This includes Swanscombe woman's 1,300 c.c. (which compares with a male brain of 1,475 c.c.), from the beginning of this period, the formerly clumsy stone tools were rapidly superseded by the pear-shaped so-called Acheulian hand-axe, a multi-purpose tool which remained dominant during the quarter million years that followed.

(c) From between 100,000 and 70,000 years ago to about 35,000 years ago, a period of between 35,000 and 65,000 years, no further significant brain expansion can be noted, except that towards the end of this period Cro-Magnon man equalled the Neanderthalers' high average of 1,650 c.c. In cultural terms it can be said that Neanderthal Man, during the relevant Mousterian sequence, laid the groundwork of most of the improved stone techniques that followed.

(d) From 35,000 years ago to the present, a period which is more or less occupied by modern man, brain capacities declined from the high average of Neanderthal–Cro-Magnon of 1,650 c.c. to a present world average of about 1,450 c.c. During this relatively short period, embracing 35,000 years, human culture expanded with ever-increasing momentum. Such culture growth includes the whole of the Upper Palaeolithic, the Mesolithic and the Neolithic, with the cultural sequence of Aurignacian, Gravettian, Solutrean and Magdalenian. It includes man's change-over from a hunter-gatherer to a food-producer, urban-literate civilization, and the whole western European cultural complex up to what might be

* The term Chellan is now rarely used and denotes here the pre-hand-axe types of stone tools.

termed the present nuclear and environmental pollution age.

The preceding considerations have been based on a comparison of average brain sizes. But this is not entirely objective, because, while the modern brain average can be compiled from a very large sample, the brain measurements of, say, the Swanscombe women have been arduously compiled from the pieces of a single brain case. Thus the specimen concerned could have been of a rare maximum brain size. But it could also have represented a minimum, or a true average. With the Australopithecines, Java, and Peking man, the situation is better, because quite a number of specimens have served as a sample. In the case of the Neanderthalers and Cro-Magnons the sample number is still larger and, while several specimens have been found exceeding 1,800 c.c., our average has been noted statistically as 1,650 c.c. In modern man there are cases of over 1,800 and even 2,000 c.c. as well as 1,000 c.c. and below; but the vast majority are around 1,450 c.c.

Thus, brain size alone does not appear to be a reliable indicator of intellectual capacities. However, if we add the compelling factor of a decrease of natural selection throughout the ages the resulting conclusions are more or less acceptable.

We may conclude then that in period (a) during man's greatest brain expansion, cultural expansion was practically nil, although the time space of 700,000 or even of a million years, must be considered enormous. Period (b), principally the age of the hand-axe, while representing a bigger step in cultural advances, must still be considered as relatively insignificant – the more so as, during the enormous period of 230,000 years which elapsed, the brain increase was as much as 350 c.c.

During period (c), one of virtual standstill in brain expansion, the early Neanderthalers laid the cultural groundwork, which in the Mousterian culture complex exhibits the incipient stages of most of the stone-work techniques which were elaborated and refined in the later Upper Palaeolithic. Hence, during a period of relative evolutionary stagnation in brain sizes (relative, because within the period there must have been many ups and downs) there was an appreciable, though not dramatic, advance in culture. During period (d), the last 35,000 years of human history, we note a recession in average brain size of as much as 200 c.c. Yet, surprisingly, this was paralleled by a cultural expansion which proceeded at an uneven yet ever increasing pace.

All this represents almost irrefutable evidence to show that cultural creativeness and expansion is not directly linked to biological evolutionary factors, except perhaps in its most incipient and crudest stages which preceded the use of fire. Thus the essential cultural advances with which we are concerned beginning with the post-Mousterian sequence and ending with our present technological civilization, were made during a period of virtual evolutionary stagnation and even recession.

6

About Evolution in Culture

It is assumed that the term 'evolution' in biology originated with Herbert Spencer to describe the progressive diversification of living forms as postulated by Darwin in *The Origin of Species*. Spencer not only applied the word to continuous biological improvement culminating in modern man, but equally to social (cultural) progress. Social (cultural) evolution was to him simply the continuation of human biological evolution. The Spencerian thesis was further elaborated by Edward Tylor, who concluded that the apparently universal belief of primitive people in a spirit world (universal animism) was generated by the similar working of the human mind (implying a biologically linked evolutionary sequence). Tylor proceeded to describe cultural progress as passing through the three stages of Savagery, Barbarism and Civilization. The American anthropologist Lewis H. Morgan (1877) postulated a social evolutionary sequence of fifteen ethnic periods starting with lower savagery and ending in civilization.[1]

One of the most complete catalogues demonstrating parallel cultural evolution was Frazer's *The Golden Bough*. He showed that most of the primitive ideas and customs held by primitive people all over the world, even the crudest and oddest ones and expressing themselves in mythology, primitive religion, totem, taboo, kingship-systems, initiation-ceremonies, etc. were so strikingly similar that they had to be explained in terms of a common factor. Frazer, following Tylor, chose as this common factor the similarity of the working of the human mind in like circumstances, the principle of the psychic unity of mankind. In other words, social institutions and customs were assumed to be the product of socio-cultural evolutionary factors representing stages in a biological/evolutionary linked sequence.

What makes the clarification of the problem of cultural evolution in anthropology so difficult is that the term evolution has become tainted with ambiguity as it is used both in a biological and in a non-biological sense. This probably results from the dictionary interpretation of the terms 'to evolve' and 'to develop' which are both described as 'acts of unrolling and unfolding' (see for example, Chamber's *Twentieth Century Dictionary of English*). Thus the term 'evolution' in culture has been used in an evolutionary biological sense by Spencer, Morgan etc. and in a non-biological sense by others.

For example, W. J. Perry, who is a known anti-evolutionist in anthropology, has written that 'warfare is not coeval with civilization; it took its intense form at a later stage of development, it is, in a way, a by-product of social evolution'.[2] Obviously Perry refers here to a non-biologically linked social development, but the uninformed reader may well think that Perry implies Spencerian evolution. In *The Stone Age of Northern Africa*, McBurney speaks of a complete evolutionary series for the hand-axe.[3] Now it is clear from the text that McBurney does not signify any biological linkage. However, in both cases a double meaning is inevitably implied. Furthermore, the word 'evolution' is now used variously to describe any kind of development, so that one may speak of the evolution of the game of chess, or of pop-music, or of football. In the latter case it means the development from a crude game kicking around an inflated animal bladder to Association Football, counting millions of adherents throughout the world. Here the use is unambiguously non-biological.

Drawing a lesson from all this it is of considerable importance to restrict the term 'evolution' in anthropology to Spencer's clearcut meaning which implies a biological linkage. This at least is the rule followed in this study.

That such a distinction is essential has also been pointed out by Nobel Laureate, P. D. Medawar, who wrote that 'Spencer was the first evolutionist and he gave the word "evolution" its modern connotation in English'.[4] The same point, though differently formulated is made by Lévi Strauss who says:

> 'The evolutionist interpretation in anthropology, clearly derives from evolutionism in biology. Western Civilization thus appears to be the most advanced expression of the evolution of societies, while primitive groups are "survivals" of earlier stages, whose logical classification reflects their order of appearance in time.'[5]

7

Culture Interpretation in Time and Space

In view of the long-standing evolutionary stagnation in the human species it can be assumed that palaeolithic man of 20,000 years ago, or of a hundred thousand years ago, or even of 300,000 years ago must have been just as competent as later neolithic man mentally and physically to start agriculture (which was not practised before 10,000 B.C.), produce pottery (which was not known before the seventh millennium B.C.), or invent writing (which was not practised before the fifth millennium B.C.). But why did palaeolithic man fail to do what neolithic man succeeded in doing after him; and why did all the above-mentioned cultural creations appear so late in human history and in a distinct historical context? The answer is: for the same reason that our ancestors living in the fifth, the tenth, or even in the fifteenth century of our era (A.D.) failed to produce a steam engine, assemble a microscope, build an aeroplane or create an amusement park like Disneyland. Each of these achievements represents the final link in a continuous chain of material (or technical) developments of great complexity and without the completion of every step in the sequence the final achievement could not have occurred.

In the case of culture creations of an institutional nature we can take the example of Christianity and ask the question, why did Christianity not appear in 1000 B.C. or in 100 B.C., or at any time before the birth of Christ? The reply is, for the same reason that Buddhism did not appear in ancient Egypt or that a fully-fledged British parliament did not assemble before the fourteenth century A.D. The reason is that the creation of Christianity was entirely due to the particular personality of Jesus Christ who appeared in a certain historical context, which was unique and which could not have been duplicated anywhere else. And the same is the case with Buddhism. In the case of the British parliament we are unable to trace its origin to the inspiration of one man, but must see it as the result of growth, the collective result of many individual actions, decisions, ideas, all arising from a unique historical development. Also when we later examine the customs of burying the dead in a crouched position, the use of red-ochre (both creations of the Mousterian culture), and the emergence of divine kingship (which made its entry into history during the early stages of Sumerian Egyptian civilization), we find that their origins are just as unique as are those of Christianity or the British parliament or other world-wide belief-systems and governmental institutions.

The origin of cultural elements should, therefore, be sought in particular sets of historical circumstances and further cultural developments attributed to their growth, spread and continuity. The prevalence of world-wide cultural traits of a similar nature can then be explained as resulting from their spread from centres of origins (i.e. cultural diffusion). To illustrate this process some specific archaeological and ethnological evidence is presented in the following:

CROUCHED (FLEXED) BURIAL AND THE USE OF RED-OCHRE

Both these cultural elements are specific creations of Neanderthal people (Mousterian culture) which lasted approximately a hundred thousand years. Oakley placed the oldest known Neanderthal remains at 120,000 B.C. (Weimar-Ehringdorf, Germany) and youngest remains at c. 20,000 years (Tafforalt, Cave, North Africa).[1] Denise Bordes[2] maintains that definite evidence of human burials is only available since Mousterian times. And Professor Francis Bordes[3] points out that most Neanderthal burials are associated with the Quina-type and Ferrassie-type Mousterian cultures, though some are from the typical Mousterian.

> We also know that in every Mousterian context mineral ochre [manganese-dioxide, red-ochre] was used. Some bits have been scratched to make coloured powder, others are pointed, like pencils. These colours presumably served for body painting as there is no cave painting that can be traced back to the Mousterian. The Mousterian custom of burying the dead became generally adopted by the Upper Palaeolithic people of western Europe.

F. Bordes says[4] that the body buried in a flexed position, was almost always sprinkled with red-ochre and adorned with necklaces and armrings of shells and pierced teeth.

An eastward spread of these customs during Upper Palaeolithic times is noticeable from an example given by B. Klima who speaks of a female resting in a flexed position with her head and chest sprinkled with red-ochre. The place is Dolni-Vestonice in eastern Czechoslovakia and the time about 28,000 B.C.[5] Still further east, many millennia later, we find crouched burials at Jericho in Palestine during Natufian times (before 6,500 B.C.).

Many examples, almost contemporary with the Natufian, can be cited from the so-called 'Fertile Crescent' of western Asia, dating from between the eighth and the fourth millennia B.C. J. Mellaart[6] has described the discovery of a cemetery of twenty-six burials at Zawi-Chemi, northern Iraq, dating from c. 8,500 B.C., including a young woman buried in a flexed position accompanied by red-ochre, a grinding stone and a necklace of small beads. Another discovery described by Mellaart[7] is that of over a hundred graves at Ell-es Sawwan (date 6,000 B.C.) in the northern Mesopotamian Plain. At least one of the burials is covered with red-ochre. Mellaart also mentions the burials of the victims of a massacre at Haji Firuz on the Iranian Plateau, dating from 5,125 B.C. where twenty individuals were found in a mass grave with red-ochre sprinkled over them.

The further path that the custom of the crouched burial and the use of red-ochre took can be traced from discoveries made in the Lake Baikal area in Siberia, described by Okladinov and dated to the Kitoi phase. He writes that 'the burials of the Kitoi phase which follow the Serevo graves in time (third and beginning of second millennia B.C.) stand out as a result of one specific feature of the mode of the burial, the custom of sprinkling red-ochre over the entire body'.[8] According to Okladinov such burials are also contemporary with the use of cord-marked pottery. Some time later this association between cord-marked pottery and the use of red-ochre turns up in north-east America, in what is now the state of New York, during the early Woodland phase. In its context, cord-marked pottery is contemporary with mortuary observances (cremations) in which red-ochre was extensively used.[9]

Turning our attention to the continent of Africa, we find that in pre-dynastic Egypt the custom of the flexed burial was widely adopted. According to Emery[10] the custom was continued during the dynastic period, and the graves of the peasants of that time (as distinct from nobles) differed little from those of the late-pre-dynastic period.

In Egypt, amongst many other uses, red-ochre turns up in the preparation of mummies. Professor Elliot Smith points out that the practice of painting the body of a mummy with a coat of red-ochre (as part of other practices) was prevalent between the nineteenth and the twenty-second dynasties. This also includes the painting of the head of the mummy in black.[11]

Dr G. L. Pretty[12] says that a similar procedure was followed in the preparations of mummies from the Torres Strait (separating New Guinea from Australia). Reviewing the evidence of the famous Macleay mummy from Darnley Island, he writes: 'The whole body was then given a coating of red-ochre. The head had been shaved and the scalp painted an even black.'![13] The date of this remarkable practice is the nineteenth century A.D. Thus we encounter an identical use of red-ochre in association with funerary practices (an unmistakable relic of Neanderthal times) in two distant parts of the world separated from each other by a time gap of about 4,000 years. And both of them are again separated from Neanderthal times by perhaps more than 40,000 years.

Some ethnologists have attributed to religious motivation the Neanderthal burial custom of placing the body in a flexed position, resting on its side and covered with red-ochre, but no valid evidence has ever been produced to support this speculation. Why, then, did Neanderthal people bury their dead in this crouched position and covered with red-ochre? We cannot answer this question, but the following considerations seem to be relevant.

Firstly, the interment of a corpse in a crouched position resting on its side is an entirely arbitrary procedure for there are dozens of other positions into which a corpse can be placed. Even in a flexed position, the corpse can be placed face-downwards, or lying on its back face upwards. Apart from this a corpse can be buried standing on its feet, or in reverse, standing on its head. It can alternatively be folded in a jack-knife fashion, lying either on its right side, on its left side, on its face, or on its back. As a matter of fact, the most usual position for burial, which has become customary during the last few centuries and even millennia and which is used by millions of peoples today, is that of the corpse stretched out on its back. It is doubtful whether any reason can be given for this practice, except that it is a continuation of what our fathers did and what their fathers did before them, none of them knowing why. Furthermore, there has been no single example of any attempt in recent times to indicate a return to a crouched burial and the sprinkling of red-ochre on a corpse. Indeed, if such an event should happen, it would arouse the greatest curiosity. The assertion that the two customs of crouched burial and the sprinkling of red-ochre, either alone or in association, could have been invented independently and spontaneously in different parts of the world must, therefore, be set aside as entirely untenable. Instead we must assume that at a unique occasion during Neanderthal times, for specific reasons which we will probably never discover, a certain group of people for reasons unknown, began to bury their dead in a crouched position and covered them with red-ochre. And if we find these same customs repeated in a world-wide setting in later ages, we must assume that they are ultimately related to the original Neanderthal practice.

POTTERY AND AGRICULTURE

Almost identical cases for the origin of countless other cultural creations and their world-wide spread from one place can be demonstrated in the same way as crouched burial and the use of red-ochre.

Pottery appears for the first time in the Fertile Crescent in western Asia; between 7,000 and 6,500 B.C. it spread throughout the thousand miles or more of the Fertile Crescent.[14] Expanding westwards it reached Yugoslavia about 4,900 B.C. and Holland about 4,250 B.C.

In Asia the earliest traces of pottery are in Thailand c. 5,600 B.C.,[15] indicating continuity of crude burnished ware with the Fertile Crescent of the seventh millennium B.C. No intermediate finds exist yet to trace the path of this likely transmission. On the other hand, continuity of cord-marked pottery (an accompaniment of the burnished ware of Thailand, 5,600 B.C.) between Thailand and China of the fifth millennium B.C. and China and Siberia of the third millennium B.C. can be established. American cord-marked and rocker-stamped pottery dating from 2,500 B.C. shows signs of continuity with Asia.

Thus we have a fairly clear-cut continuity reaching from western Asia to America, supplying a list of successive dates all falling within a space of 4,500 years. In spite of this obvious picture of interdependence, various claims for independent pottery origins are still being made.

There are two factors which militate against independent origins of pottery in more than one location of the globe. One is the complex nature of even the simplest form of pottery-making, and the other is the world-wide appearance of pottery and its measure of geographical continuity, all compressed within a very short space of man's total history.

Gordon Childe has pointed out that ...

> ... in its crudest and most generalized form, pottery itself and the craft of the potter were already full of complexities, and that pottery involved a number of distinct processes and the application of a whole constellation of discoveries.

The potter had to learn how to find and to choose the proper clay, how to clean it, with what proportions of water and grit it should be mixed, etc. He adds:

> The shaping of the pot itself is not so easy as it sounds. Quite small vessels can, of course, be kneaded and moulded, mudpie fashion, out of a lump of clay. Or a coating of clay can be spread over an open

basket or a half-gourd; when it has dried, the form can be removed, and you have an open dish or platter ready for firing.

But if anything larger is desired, or a vessel with a narrow neck like a bottle or jug, such elementary processes no longer suffice; the vessel must be built up. In neolithic Europe and Asia this was generally done by the ring method; after the base had been moulded, rings of clay of the desired diameter were prepared. One was attached to the base and then another set on the top edge of this, and so on. It is a slow process. The rings must be fairly wet and plastic when they are put on. But as soon as one ring is in place, you must pause and let it dry and harden but not too much before adding the next storey. The mere construction of a large pot may take several days.[16]

We can see that such a complicated technique as pottery-making is unlikely to have been independently discovered in more than one geographical location and during more than one particular historical period. As a matter of fact, the archaeological record shows that (at the time of writing this) from nowhere in the world have traces of pottery been reported preceding those found in the Fertile Crescent of western Asia and Thailand during the seventh millennium B.C.

All the known basic pottery types of the world made their appearance within a period of about 4,500 years and within this period, in all cases where sufficient evidence is available, interdependence and continuity can be established.

AGRICULTURE

The earliest known evidence of plant cultivation relates to three places, all in the area of the Fertile Crescent, and all dating from about 7,000 B.C. They are Hacilar in western Anatolia (Turkey), Beidha (Jordania) and Ali Kosh (Khuzistan, near Susa).[17] What strikes us most about these early agricultural finds is, firstly, the closeness of the dates of their origins, and secondly their location within a restricted and altogether small area of the globe. Moreover, all later traces of agriculture in this area show interrelationship and continuity and fit into the time and space boundaries defined by archaeological evidence.

What remains obscure is the location of a first centre of origin, preceding the three places mentioned and possibly pre-dating them by several hundred years. Another alternative could be that one of the three places itself may yet reveal earlier origins than hitherto assumed. Agricultural continuity in time and space can also be established between western Asia, Europe and Africa. All the early agricultural settlements that have been found there are part of what we may call a barley–wheat–burnished-pottery complex of great similarity. In eastern Asia, in contrast, we can speak of a rice–burnished and cord-marked pottery complex, which on present evidence shows no direct continuity with western Asia, and finally, there is a maize–cord-marked and rocker-stamped pottery complex typical of America, which shows traces of continuity with Asia.

However, while there are many missing links in the chain of agricultural continuity as it extends in the general direction of western Asia–Thailand–China–Bering Strait–North- Middle- and South-America.[18] There are also many indications which point at possible contacts across existing gaps. For example, in the case of Thailand, Gorman[19] mentions the earliest traces of cord-marked pottery and burnished ware in association with rice. The dating of pottery (c. 5,600 B.C.) indicates continuity with the burnished ware –barley–wheat complex of western Asia (which is 1,000 years older) over a route which is as yet unknown.

As to a continuity between Far Eastern Asia and America, we can recall the remarkable coincidence of rocker-stamped Jomon pottery appearing in Japan c. 3,000 B.C., and in America during the Valdivia phase of Ecuador at c. 2,500 B.C. Continuity can also be shown between Siberian cord-marked pottery and Woodland cord-marked pottery in north-eastern America in connection with red-ochre. Both are later associated with agriculture. However, the latter evidence is of such a late dating that it cannot be used to explain the much earlier appearance of agriculture within the New World itself. However, what Chang said about eastern- and south-eastern Asia[20] certainly applies to America also; namely, considering that there was north-south contact in pottery technique and style, there could have been no effective barrier against the transmission of ideas for the cultivation and use of plants either.

Those archaeologists who insist that agriculture has been invented independently and in isolation in western Asia, Europe, Eastern Asia, America and possibly at several other places, are certainly disregarding the evidence. New evidence dug out from the past points increasingly in the opposite direction, the diffusion from a single source of origin. This assessment holds good even without taking into account the decisive factor of a long-standing arrest in human biological development. At the same time it has to be admitted that in the special case of agriculture, due to the perishable nature of the product, it is in many cases difficult to supply direct proof for diffusion and continuity. Nevertheless, there is an increasing tendency amongst scholars to substitute diffusionist factors for formerly held autochthonous assumptions.

In summing up, it can be repeated that all the known basic pottery types of the world, as well as early agricultural traces, made their appearance in western Asia, Europe, Africa, Eastern Asia, and America within a time space of about 4,500 years (between 7,000 B.C. and 2,500 B.C.) and that in all cases where sufficient evidence is available, interdependence and continuity can be presumed.

8

Towards a Historical Conception of Culture

Glyn Daniel in his book, *The First Civilizations*, lists seven principal civilizations – Sumer, Egypt, Indus, China (Shang) and America (Olmec, Inca, Maya). He maintains that each of them is of independent origin. This thesis has now become rather questionable and it can be assumed instead that these civilizations, which show many similarities, are culturally interrelated, and that their origins (with the exception of the first) are due to outside influences or stimuli. This is not to deny that their subsequent development may have largely been of their own making.

In a chart supplied by Glyn Daniel, the beginnings of Sumer and Egypt are both marked at about 3,200 B.C.; Indian beginnings stand at about 2,500 B.C.; Chinese beginnings at about 2,000 B.C. and American beginnings (Olmec) at about 1,200 B.C.[1] Thus, according to Glyn Daniel, the world's principal early civilizations occurred within a time space of about 2,000 years. We find not only that the Egyptian and Sumerian civilizations arose almost simultaneously, but also that the beginnings of the Sumerian and Indian civilizations are separated by only about 700 years, Indian and Chinese by only about 300 years, and Chinese from American by about 1,200 years. This leaves us with reasonable time gaps, allowing the bridging of the geographical distances by migration or trading activities. At the same time, the chronological sequences also indicate the direction in which the elements of Sumerian/Egyptian civilization moved. An exception is the emergence of the first American civilization of the Olmecs, where it was recently discovered that its inception was due to direct sea-faring expeditions from Egypt, organized under Rameses III, at about 1,200 B.C. However, shortly after the Egyptian arrival in America, we also find the appearance of Chinese influences.[2]

Now human organic evolution was generally at a standstill during all the period of 500,000 years which roughly covers modern man's existence; while the 2,000 years which embody the appearance of the five first civilizations fall into a period of biological recession. The point is that while modern man was intellectually capable of creating urban literate civilizations almost at any time in his half a million years history, he did not do so until about 3,200 B.C. And then he did it relatively in a hurry, by creating comparable civilizations in five different areas of the world, all within a space of 2,000 years.

Gordon Childe, in *Man Makes Himself* wrote: 'It must not be imagined that at a given moment in the world's history a trumpet was blown in heaven, and every hunter from China to Peru thereupon flung away his weapons and traps and started planting wheat or rice or maize and breeding pigs, sheep and turkeys'.[3] Yet, this is what Glyn Daniel and other like-minded ethnologists imply when they speak of the independent occurrence of pottery and agriculture in different parts of the world within a space of 6,000 years, and of the five urban literate civilizations within a space of 2,000 years.

The only reasonable alternative to explain the occurrence of similar cultural elements in different parts of the world is to consider them derived from comparable cultures in other parts of the world which preceded them in time. Due to the incompleteness of the archaeological record, the tracing of such cultural linkages is often difficult and remains frequently unexplained. But whenever such gaps in the archaeological record occur, cultural interrelationship must be postulated, just as Darwin had to postulate the interrelationship of species. Meanwhile the gaps must be left until archaeology supplies the missing evidence.

Making his case for the independent origin of the first civilizations, Glyn Daniel writes '... I would prefer to say that civilization and history began seven times. And why? Because the societies in question met certain cultural possibilities, accepted their challenge and developed them into civilizations.'[4]

Mr Daniel also maintains, that '*there is a supra-organic, a cultural evolution in man's development*' (my italics). He quotes Kroeber and Caldwell in support. Kroeber wrote in 1940. 'We must consider that civilization is an inevitable response to laws governing the growth of culture and controlling the man–culture relationship', and in his Preface to a book of essays entitled *New Paths to Yesterday*, Professor Caldwell says: 'Perhaps there is only a *finite number* of social and historical processes behind the events of history.' 'I think,' continues Glyn Daniels, 'Kroeber and Caldwell are right; the lesson of archaeology at the present day is, I suggest, that seven societies in seven different ways trod one of these paths, the path that led to civilization.'[5]

Glyn Daniel further points out that '... the diffusionist and certainly the hyper-diffusionist model of the past is not now suitable, and the same applies to the unilinear evolutionary model of thought', and he finally concludes 'I believe that an interpretation of the origins of civilization in terms of multilinear evolution is in accordance with the archaeological facts as now known to us.'[6]

In examining the implications of these statements it is profitable to turn back to Glyn Daniel's own explanation of unilinear evolution, which he rejects. Quoting Adolf Bastian (a German scholar of the last century) as a textbook example, he writes (my italics):

'He argued that by a general law the psychical unity of man everywhere produced similar ideas; different geographical environments might produce different responses, and there could of course be contacts in later historic times, but the basic idea was the psychic unity of man and, therefore, the independent evolution of culture. These ideas of Bastian's were *a form of super-organic or cultural social evolution*.'[7]

How does Glyn Daniel's 'Multilinear evolutionary process' compare with Bastian's 'unilinear evolutionary process'? It is contended that in the latter case all societies pass through the same cultural evolutionary sequence, while in the former case only a limited number do so. For example, Glyn Daniel, citing Steward, writes:

Steward propounds a multilinear evolutionary theory according to which all peoples do not pass automatically through similar stages, but a finite number of parallel evolutions take place – not natural laws but regularities of limited range.[8]

But do unilinear and multilinear evolutionary developments, or rather their underlying causes, really differ? Not in Mr Daniel's argument. Because, as noted, he explains Bastian's unilinear evolution as '*a form of super-organic or cultural social evolution*'[9] and his own multilinear evolution as '*a supra-organic, a cultural evolution in man's development*'.[10] Indeed, a striking conformity.

How does supra-organic, multilinear evolution proceed? The clue is supplied by Glyn Daniel himself when he discusses the origin of Sumerian Civilization.

The genius of the people must be seen against the background of the possibilities of their geographical environment ... It may have been something special in their make-up, and I can not do better than end this chapter with a sentence from Kramer: 'The psychological factor responsible to no little extent for both the material and cultural achievements of the Sumerians was an all pervading and deeply ingrained drive for pre-eminence and prestige, for victory and success.[11]

And in explaining the rise of the Indus civilization, Glyn Daniel expresses similar ideas. He quotes Wheeler in support. 'A society determined to profit by the vast opportunities of the plain must needs have also the genius and the skill to master an exciting and minatory environment, and must have it from the outset.'[12]

Far be it from us to accuse Kramer, Wheeler or Glyn Daniel of any racialist tendencies. On the contrary the latter has strongly criticized such cultural views. And yet basically seen the above ideas on cultural origins are built on the same premise as was Hitler's élitist myth of Aryan cultural and racial superiority, namely, the existence of an inbred biological evolutionary factor. This special racial property, if we pursue Glyn Daniel's argument to its logical conclusions, must have been present seven times and in each case must have led to each of the seven first civilizations.

And why must this be so: because, as Glyn Daniel himself points out, identical conditions which could have led to civilization frequently presented themselves. But, apart from seven occasions, the people living under these favourable circumstances lacked some mystical quality which would have allowed them to be exploited.

'We must not delude ourselves,' writes Glyn Daniel, 'into thinking that all one had to do is to produce possibilities and they will automatically be accepted and used ... There were many possibilities existing in the world that did not in fact produce civilizations ... The genius of the people ... etc.'[13]

It must by now be obvious that the argument of Glyn Daniel for a special genius to explain the origin of Sumerian civilization will not do. Neither is there any substance in the assertion that something like a multilinear cultural evolution can have taken place to account for the independent origin of the six other civilizations. The facts of genetic evolution can simply not be bent to fit Glyn Daniel's or any other form of cultural evolutionary theory.

It is for quite some time now (principally as a reaction to Hitler's racialist outrages) that biologists have refuted the notion that selected human groups or races can be, or can have been, endowed with characteristics which make them biologically superior to other racial, cultural, or national groups. Geniuses like Aristotle, Archimedes, Leonardo da Vinci, Newton and Einstein, to name a few, may occasionally appear at any age and among any national or racial group if conditions are favourable. It can also be shown that any of the existing races and nations (and we count six major races and over 120 nations now) are able to produce doctors, statesmen, administrators, university professors, technicians, scientists, writers, artists, poets, and sportsmen, of reasonable capacities. It all depends on suitable educational and training facilities. And this applies also to people who until recently lived a very primitive life like the Kalmuck and Samojedes of Siberia or the aborigines of Australia and New Guinea. A good point in this respect was made some time ago by Margaret Mead when she related some of her experiences on New Guinea. She said: 'I have seen people, who have moved from the stone age into the present in thirty years, kids who say "My father was a cannibal, but I am going to be a doctor."' It can be concluded that the evolutionary explanation can certainly not account for the origin of the first civilizations.

According to genetical evolutionary theory as shown in the main argument of this book, organic evolution is produced by natural selection acting on fortuitously produced gene mutations. In rare cases these mutations will produce evolutionary changes. But normally even an intense natural selection will merely keep species on the same evolutionary level. In the human species with the advance of culture natural selection was reduced and biologically lesser endowed humans were able to survive and propagate. As a result no further evolutionary advance was accomplished. To speak therefore of an evolutionary process that can produce progressive cultural changes particularly during the last 10,000 years, the period which includes the invention of agriculture up to our own present nuclear and environmental pollution age, is at variance with biological evolutionary facts. On the contrary, during the 10,000 years in question, as shown by the analysis of comparative brain sizes of fossil men (Table 1 p. 60), we must not only assume the prevalence of evolutionary stagnation but even a certain amount of biological evolutionary recession.

Sumerian civilization made its appearance in the latter part of this period during the fourth millennium B.C. and we agree with Mr Daniel that the Sumerians were perhaps the most culturally advanced people of their time. But being culturally the most adept, they were also most affected by a diminished natural selection. An intensified agriculture improved their living standards and made their lives not only more comfortable than that of other contemporary human groups of a lower culture, but also made the Sumerians more susceptible to biological recession. Therefore in accounting for the cultural achievements of the Sumerians, we can say that to whatever influences these were due, they certainly did not result from evolutionary factors, whether named multilinear or nonlinear. As a matter of fact, it can be reaffirmed that during the last few hundred thousand years no evolutionary agency was active in the human species which could have produced advances in culture.

Therefore, a confrontation of the absent factor of 'evolution' and the present factor of diffusion, implied in Glyn Daniel's phrase 'evolution versus diffusion' is not feasible. Equally obsolete is the saying of Lowie (frequently quoted by Glyn Daniel) that *evolution can lie down amiably beside diffusion.*[14] In this sense, diffusion in absence of the evolutionary factor, has been and remains a lonely bed-fellow. Nevertheless, it must be added that although no evolutionary factor existed during the last ten millennia which could have created culture, diffusion was only one of several factors promoting it. Another was the historical factor. This means that in each case of cultural origins there are specific historical circumstances at work which account for their first appearance, whereas diffusion itself represents a historical process of another type. It alludes to the spread of a specific culture after its origin. Finally, there is also the factor of continuity, which accounts for the development, growth, and diversification, of already existing cultural traits, acting either at the place of origin or after their diffusion to another environment.

In the foregoing argument it was shown that cultural origins can not be attributed to the biologically predisposed 'genius' of a people. But if there is no such attribute and if only one particular group of people took advantage of a particular environment and originated a specific culture, although similar conditions presented themselves to other groups who somehow remained indifferent to the challenge of their environment, then there must be other forces at work.

Thus we have different human groups which are biologically identical, living in many cases in environments which are also similar. What can, however, not be identical and distinguishes one group from the others are their different histories. The history of each human group is unique and depends on many diverse factors which in their sum total cannot be duplicated by any other group. Therefore, the only explanation for a particular group having originated or acquired a specific culture while others did not, is the prevalence of historical accident. And if we find cultural traits and even whole civilizations elsewhere in the world which are similar, it can be assumed that they are derived from comparable cultures and civilizations who preceded them. For all similar cultures and civilizations there must also ultimately exist an original source of specific historical origin. There is, of course, a theoretical possibility that similar cultural traits can have been independently created. But the frequency of such an occurrence can be judged to be minimal. It can be likened to the chances of a blind-folded man shooting an arrow into the sky and accidentally hitting a bird. In other words, the possibility of two similar cultural elements or civilizations occurring independently in different parts of the world is almost nil, while the possibility of seven similar civilizations occurring independently in seven different parts of the world is absolutely nil.

A good illustration of how historical circumstances operate is the origin of the culture complex of Christianity. It is due to unique historical circumstances surrounding the birth and life of Jesus Christ in Palestine during the time of the Roman Emperor Augustus. In the course of the next 2,000 years, Christianity spread world wide. And if today we find the cult of Jesus on the cross and that of the virgin Mary in the remotest nooks and most isolated corners of the globe, we can be sure that its existence is due to a process of world-wide diffusion, originating in the then Roman protectorate (province) of Judea. In no instance can the prevalence of the Jesus cult in different parts of the world be attributed to independent origins.

Another example of the historical process, visible this time not in the creation of an entirely new culture but in the diffusion of an already existing culture, is that given by Mr R. A. Jairazbhoy in *The Egyptian Origins of the Olmecs*. It documents the historical circumstances which around 1,200 B.C. led to the diffusion of Egyptian cultural elements into Mexico and initiated the first American civilization.[15]

Yet the most topical example for seeing the historical factor at work is our own contemporary world. Because, although the more than 3,000 million people living today are for all practical purposes biologically uniform, and although many of the societies they form live in very similar natural environments, they show a meddley of cultures of dazzling variety. These range from the most advanced nuclear-age civilization of the United States to a group like the recently discovered stone-age tribe of the Tazaday in the Philippines, who before their discovery had no knowledge of agriculture or metals, and had never tasted salt, sugar, or tobacco. But the United States civilization itself is again a medley of cultures. There are still Red Indian tribes following pre-Columban customs, and countless immigrant groups continuing to practice the cultures of their home countries. Most of the inhabitants of the very modern society of Israel regulate their life by religious practices which were prevalent more than 4,000 years ago.

Within the Babel of cultures inhabiting the world we can observe that an individual person's culture is determined by the society into which he is born, while the culture of the society itself is due to the accidental working of historical circumstances. Yet, as we go back in time it becomes more difficult to reconstruct the exact historical circumstances which must have led to the origin of specific cultural elements. In most cases no written record exists at all and past events have to be pieced together from scattered archaeological remains. Needless to say that to fill the gaps, the historical explanation involves much speculation. But to whatever such attempts of historical restoration amount, the true history must in each case be unique.

Let us recall some of the archaeological evidence relating to the origin and spread (diffusion) of such basic elements as a specialized food collecting economy, the cultivation of cereals, the making of pottery, and the origin of whole civilizations. In the cases of food collection, cereal cultivation, and pottery, it is doubtful whether the exact historical circumstances of their origins will ever be discovered. The situation for the first civilizations is more hopeful. We can, however, show that in all cases of first appearances on the historical scene the rapid appearance of similar cultural elements in many other parts of the world follows. It is also important to repeat that all these events should be seen against the background of 500,000 years of history; a period during which – on countless occasions – biologically equally endowed human groups inhabited environments which were amenable to a specialized food collecting economy (cereals), to the actual cultivation of cereals, and to the production of pottery. But it was only towards the very end of this long period, at c. 11,000 B.C., a specialized food collecting economy made a first appearance in the area of the Fertile Crescent. Such evidence comes from Natufian Palestine and from Zawi Chemi in the northern Zagros and is in both cases of equal antiquity. About 2,000 years later we see similar activities emerging in Thailand (c. 9,000 B.C.) and another 1,000 years later in America (the desert cultures of Mexico, c. 8,000 B.C.). If in the face of future discoveries, the priority of the Fertile Crescent in this respect should remain unshaken, we can postulate that unique historical events happening there were responsible (and, for that matter, the same would apply to any other area of origin). Other similar economies of a later date, like those in Thailand and Mexico, obviously derived from this source. Such a process must apply whether or not we will ever know the exact historical details which led to these cultural origins, or the manner and direction of their spread.

It was not until c. 7,000 B.C. (i.e. 4,000 years later) that the next step was taken which led from mere cereal collection to deliberate cereal cultivation. This appears to have again happened in the Fertile Crescent. And although originally only one specific human group can have been responsible, its location has not yet been determined. Instead, we find three places where cereal cultivation occurred almost simultaneously at about 7,000 B.C. They are Halicar in Turkey, Beidha in Jordan, and Ali Kosh to the north of lower Mesopotamia. According to our theory we must assume that either all these three places derived knowledge from a fourth as yet unknown place, which preceded them in time, or that one of the three was the first which subsequently inspired the others. Be that as it may, what further happened is that during the next 1,000 years cereal cultivation spread over the entire area of the Fertile Crescent, including Jericho, Jarmo, and Jeitun. Of almost equal antiquity are finds in Thailand (rice cultivation at Spirit Cave). In a more westerly direction places of early cereal cultivation include Cyprus and Greece (c. 6,000 B.C.) and the Netherlands (c. 4,200 B.C.). In China cereal cultivation (rice and millet) appear from c. 5,000 B.C. onwards and in Formosa from c. 4,000 B.C. The first evidence of cereal cultivation (maize) in America has variously been indicated as occurring in Mexico between 5,000 B.C. and 4,000 B.C. If all these events are pitted against the background of the last half million years of human history we find that the time intervals which elapsed between the first appearance of the aforementioned cultural elements and their world-wide spread are minimal. The world-wide spread of specialized food collection took barely 4,000 years (11,000 B.C. to 8,000 B.C.), that of cereal cultivation c. 3,000 years (7,000 B.C. to 4,000 B.C.) that of pottery c. 4,500 years (7,000 B.C. to 2,500 B.C.); and that of the first five civilizations Sumer, Egypt, India, China and America (Olmec) c. 2,000 years (3,200 B.C. to 1,200 B.C.). The preceding figures rest on presently available evidence. This means that we must be prepared to amend these time sequences should evidence be discovered which pre-dates present finds. But even if such amendments in each case amounted to several thousand years, it would not matter. Because seen against the background of 500,000 years of human history presenting the same possibilities such corrections are insignificant.

SOME FUNCTIONALIST VIEWS:

Some anthropologists and sociologists of the functionalist school maintain that a society will develop a certain culture because it fulfils a useful function and is necessary for survival. This presupposes that there is a biological factor at work (i.e. an evolutionary factor) so that any group on reaching a certain state in its biological development (evolution) and living in a certain natural environment, will react culturally in the same way as other groups having reached the same biological stage and living under similar environmental conditions. In this manner similar cultures and civilizations are explained as arising independently out of pure necessity. Hence, at a certain stage of biological evolution, fire and protective clothing would have come into use and at other stages, agriculture, pottery and metal-working.

However, the rationale of restricted time sequences relating to such cultural traits as intensified food collection, cereal cultivation, and pottery, disallows functionalist assumptions. Because if the genesis of culture operated on the basis of necessity or useful function, the preceding time sequences, embodying specific culture creations and their spread, could not have kept within such narrow time limits. Instead, countless similar cultures would have appeared haphazardly at any time throughout the last few hundred thousand years. Because, by then all mankind had already reached the same biological evolutionary stage as now, and similar environmental conditions were repeatedly encountered.

But there are also many documented historical examples which testify of the absence of functionalist influences. For example, it is believed that the art of making fire has been known since Peking Man (c. 500,000 years ago). One should therefore assume that all presently living societies should automatically know how to make fire. But this is not so. Daryl Forde writes that the Boro and other Japura people of Brazil are today ignorant of any method of fire-making.[16] The use of protective clothing which fulfils another vital cultural function is perhaps as old as fire-making. Yet when Darwin encountered the Fuegians of Patagonia (South America) in the last century, they ran naked in a climate that is particularly harsh and inhospitable. According to Forde, the Kwakiutl of British Columbia, in spite of a cold (near Arctic) climate, had no footwear and were running barefoot.[17] Pottery making is assumed to be a common world wide art, known since c. 7,000 B.C. yet the Blackfoot Indians of central North America had no pottery and were using fresh pounches and hide for cooking purposes.[18]

There are also many examples of peoples who have culturally stagnated for thousands of years (see the example of the Tasaday) or who have lost cultural elements they formerly possessed. Forde points out that 'rice or millet cultivation may have been known by some of the ancestral Polynesian groups but that it was later abandoned. This would be no more surprising than their probable abandonment of both pot-making and weaving.'[19] The same author also reports that the Semang, a food gathering people of Malaya, in their everyday pursuits depend greatly on stone tools, which are nothing but rough shapeless stones. These are picked up when needed and then thrown away again. Hence, it can be contended that the Semang are perhaps culturally more backward (though biologically certainly not inferior) to the people who made and used Acheulian stone tools about 3 million years ago.

These examples certainly contradict the view that evolutionary or functional processes can play any decisive part in the creation and dissemination of culture.

All this must be kept in mind when trying to reconstruct the historical circumstances of the origin and spread of the early civilizations. According to Glyn Daniel (*The First Civilizations*), the first fully fledged civilization was that of the Sumerians, emerging about 3,500 B.C. to 3,200 B.C., and that Egyptian civilization which was of almost contemporary origin was an independent creation. Both had agricultural economies depending on irrigation, both used the same calendar, had some form of writing, and they were both overlorded by kings and both worshipped gods. It should be stressed that on available evidence none of these rather elaborate cultural elements existed anywhere else in the world before 3,000 B.C. That the complicated cultural features listed above can have occurred almost simultaneously and independently both in Sumer and Egypt is hardly likely considering that not even such modest arts as fire-making, the use of clothing, and the manufacture of simple Acheulian stone axes (of 300,000 years ago) have ever been universally adopted. Also, the priority of Sumerian civilization over Egypt, which has been frequently asserted, is by no means conclusively proved. Not that it matters really for our argument. What is more likely, is that many cultural elements found in both Sumer and Egypt, had their origin in either one or the other place. Perhaps the most elemental feature common to both, was irrigation. According to Mellaart, irrigation was first introduced into the alluvial lowlands of Lower Mesopotamia c. 5,500 B.C.[20] No details as to how are available. But a clue is given by the cultivation of linseed, which needs extensive watering characteristic of irrigation, and which was grown in the Hassunah area c. 5,700 B.C. before the opening up of the lowlands.[21] Mellaart also mentions that Lower Mesopotamia (i.e. the area of Al-Ubaid, Eridu, and Ur) being situated outside the rain belt, was shunned by farmers until they had somehow acquired the knowledge of irrigation. It can therefore be presumed that those Sumerians who were the first inhabitants of this formerly inhospitable area made it their home only after having mastered irrigation techniques in their former abodes at the foot of the Zagross mountains, probably in Khusistan.[22] As it is fairly certain that cereal cultivation and animal rearing had been previously introduced into Egypt from the Fertile Crescent even before irrigation was known in either place, it can be speculated that late-comers from Lower Mesopotamia (i.e. the Sumerians) subsequently also brought the knowledge of irrigation to Egypt. Glyn Daniel writes that

... both the Sumerians and the Egyptians created civilizations based on irrigation. But it would be a mistake to conceptualize irrigation. Instead it is a very different thing in Mesopotamia and Egypt. In the land of the twin rivers it was a process of making canals, there were canal banks above the fields, the water drained away on to the fields, and sank into the soil, and this is why, in due course, the south Mesopotamian fields had to be abandoned because of salination. This never happened in Egypt; Egypt was the gift of the Nile in the sense that the Nile did do all the work: it all happened with the flooding of the Nile. In Egypt the mud dries out but the salt remains deep in the cracks and drains from the field to the river.[23]

This difference in irrigation technique between Sumer and Egypt is of considerable consequence, because while the Nile irrigation was entirely dependent on the natural flood water, a regular annual event, Lower Mesopotamia irrigation was an arbitrary operation depending on the less regular opening and closing of irrigation channels. Hence, in Sumer, irrigation was adapted to the progress of the crops, while in Egypt the crops had to be adapted to irrigation. In comparing the two methods it becomes obvious that for the Egyptians the use of a calendar, for predicting the annual arrival of the Nile flood, was of vital importance, while in Mesopotamia a calendar was not so essential. The general assumption that the Egyptians were the first to invent a calendar based on the annual flood cycle of the Nile, is therefore well founded and this is supported by archaeological evidence. Gordon Childe points out that to predict the flood the Egyptians had to determine the length of the solar year in days and devise an artificial calendar to reconcile solar and lunar years.

Now it happened that just about the time when the flood should reach Cairo, the last star to appear on the horizon, before dawn obscures all stars, was Sirius, the Egyptians' Sothis. The helical rising of Sirius was thus a naturally fixed point in the solar years. It was found that this event recurred roughly every 365 days, and it was then taken as the starting point of an artificial state year of that length.[24]

The Egyptian calendar was accordingly divided into twelve months of thirty days each plus five days extra and an additional day every four years. The fact that the same calendar came also into use in Sumer, bears evidence that cultural fertilization of the two areas was not a one way traffic. But the significance of large-scale irrigation and of the calendar goes far beyond mere time-measuring. It seems also to be connected with the first appearance of kingly rule and the worship of gods and religion. Glyn Daniel writes about Sumer: 'Irrigation and drainage involved complicated and cooperative efforts requiring control, organization, and a centralized society.'[25] A similar sentiment is expressed by Gordon Childe about Egypt. He points out that 'the invention of the solar calendar making possible the accurate prediction of the flood's advent would be a sure way of justifying such pretensions and consolidating such authority. To enforce them, the actual power of cutting off real water by blocking irrigation channels would be an effective means.'[26] There can be little doubt that the need for such central control, possibly occurring in Sumer before Egypt, will have led to the origin of kingship. As we have already indicated, it can be assumed that irrigation started in Sumer sometime during the Al-Ubaid phase, leading to a centralized administration headed by a king. If we assume that the Sumerians brought irrigation to the Nile Valley more than a thousand years later, it is most likely that they introduced the institution of kingship as well. However, Sumerian kingship was and remained a terrestrial affair, while in Egypt from a very early stage onwards, kings were endowed with the supernatural prerogatives of gods (divine kingship). But as there is no conclusive evidence of the existence of gods and their worship in Sumer before the Ubaid period (c. 5,000 B.C.) and as it can be assumed (although the argument for lack of space cannot be presented here)[27] that religion and the worship of gods was a direct consequence of divine kingship, it may also be assumed that the Egyptians introduced gods and religion into Sumer.[28] Glyn Daniel points out that 'In Egypt the ruler was a god in his own person: he was not the human agent of the god – the Pharoh is the classic example of the god incarnate as king.'[29]

When we examine the diffusion of kingship, which is world-wide, we find that it was not the earth-bound type of Sumerian kingship which made an appearance in many of the civilizations which followed, but the more spectacular type of divine kingship. One thing which is sure is that Egyptian divine kingship and sun-worship had a profound and lasting effect on the civilizations of the world. The direction and manner of their spread has been well documented with numerous examples by J. W. Perry in his major work *The Children of the Sun*. And although some of Perry's conclusions and speculations have not stood the test of time, the basic principle of Egyptian precedence in this respect seems to be well founded.

In view of the long-standing human evolutionary stagnation, all this sounds like common sense. But many established archaeological and anthropological cultural theories, suffering from a Lamarck-based evolutionary hang-over, stubbornly reject the diffusionist analysis of world-wide cultural spread from centres of origin and instead postulate many independent origins. The main argument advanced is that there is no historical evidence to prove world-wide diffusion. This has now been contradicted by the discoveries of Mr Jairazbhoy regarding the Egyptian origin of the Olmec civilization in America. This represents a decisive breakthrough for the historical and the diffusionist thesis. It also raises the hope that other historical evidence will be discovered to document other world-wide cultural ties. Meanwhile, of course, we have to rely on indirect evidence. Such evidence rests on the fact of the world-wide distribution of *homo sapiens*, which must have been accomplished across vast expanses of water long before Columbus made his historic voyage to the Americas.

The resourcefulness and the skill of early men has hitherto been greatly underestimated. For example, recent archaeological evidence shows that Australia was occupied by stone-age people at least since 18,000 B.C. and perhaps even before. D. J. Mulvaney writes that the so-called Keilor skull, which is judged to be 18,000 years old and which was found near Melbourne, is nearly a twin of a fossil skull of undetermined age from Wadjak in Java, a distance of more than 2,000 miles away.[30] The same author also mentions a similar skull from the Niah Caves in Sarawak (north of Java), which is c. 40,000 years old and which also indicates Australian linkage. Other evidence of trans-Indonesian and even of Indo-Chinese linkages with Australia is also mentioned. This refers to saw-edged stone flakes of c. 6,000 B.C. which are comparable to similar ones of the south-east Asian Hoabhinian pebble industry.

What all this suggests is that since 18,000 B.C., and even before, south-east Asian people must have crossed the large distances separating them from Australia. And although there may have been some connecting island bridges between various of the Indonesian islands which have now disappeared,[31] these early people of the Upper Palaeolithic must have been able to make long sea-voyages. Because, how else could they have reached the places where their remains are now being found? Such sea voyages must have continued on an even larger scale in the latter stages of the pre-Christian era, as the distribution of the Malaysian language shows. It spreads over the whole of Indonesia and embraces many of the islands of the western Pacific as far as the Society Islands. These movements of early people also suggest that the intensified and specialized food collecting economy of Thailand, dating from c. 9,000 B.C. can have been transferred directly by sea voyage to the eastern Pacific shores of North America, thus introducing their similar pre-agricultural activities exemplified in the desert cultures of south-western North America, (8,000–5,000 B.C.).

However, direct evidence of long sea voyages in the Pacific eludes us until the historically documented sea voyages of the Polynesians. Daryl Forde writes:

> The Polynesian colonization of the central Pacific probably began in the early centuries of the Christian era, and there were several periods of extensive voyaging. It has been estimated, from the comparative study of genealogical records and of cultural distribution, that the sacred arri families of central Polynesia arrived from the West little more than a thousand years ago. That they introduced many new customs and modified many existing practices is claimed from the cultural distinction between arri and commoners that existed in Tonga, Samoa and particularly in Tahiti in recent times.[32]

Forde also speaks of the large ocean-going vessels of the Polynesians more than a hundred feet long and five feet deep, which could carry several hundred people with provisions and water.

> Equipped with these boats the Polynesians covered millions of square miles of sea in the space of a few centuries. The Society Islands were one of the great centres of expansion. From there repeated journeys were made to Hawaii and New Zealand during the centuries following the arrival of the arri. In the great voyages from Tahiti to New Zealand the navigators ran down to Raratonga in the Cook group, and from there launched out on a ten to twenty day voyage of about 1,600 miles, which was made in December, when the wind was favourable. The return could best be undertaken in June.[33]

Turning back to the early voyages of the stone-age people before the tenth millennium B.C. of which there is no direct evidence, but which are nevertheless indicated by the remains of these people in Asia, Australia, and elsewhere, we may say that later Neolithic peoples must have taken over these seafaring skills and that these were further continued by the early civilizations of Sumer and Egypt. Representations of ships which were in use during the early dynastic periods in both these civilizations have been preserved. Looking at a world map it can be seen that for Egyptian ships setting out from the Red Sea, it should have been easier to reach the Indus Delta than the mouth of the Euphrates. Therefore Egyptian divine kingship, reaching Sumeria via the Persian Gulf, can have exercised direct civilizing influences on India and even on Malaya and Indonesia (which are double the distance away by sea) without touching Mesopotamia first. However, all this is conjecture.

To demonstrate the extent of early diffusion there is one significant cultural parallel of a different order which has puzzled archaeologists for some time and which shows the spread of a distinct cultural element over an expanse of water covering

13,000 miles. This is the appearance in Ecuador, South America, of the highly specialized Jomon pottery of Japan. 'Betty Meggars, Clifford Evans and Emilio Estrada have emphasized the appearance of this pottery on the coast of Ecuador which they find exactly paralleled in the Jomon Neolithic of Japan.'[34]

The Jomon pottery dates from c. 3,000 B.C. onwards and the similar Valdivia pottery of Ecuador dates from c. 2,500 B.C. As there are no traces showing that this highly specialized pottery made a long journey from Japan via the Aleutian Islands or the Bering Strait into America and then all the way down the North American continent to South America and Ecuador, the only sound explanation is that transmission occurred by sea across the Pacific.

Other historically documented sea voyages which are worth mentioning here are those undertaken during the reign of Rameses III at c. 1,200 B.C. from Egypt to Mexico, recently reported by R. A. Jairazbhoy; the circumnavigation of Africa under Pharaoh Necho, c. 600 B.C. (reported by Herodotus); the voyages of the Vikings from Norway to Iceland, south Greenland and Newfoundland between A.D 800 and A.D. 1200; Marco Polo's return sea voyage from China to Persia at c. A.D. 1291. Those sea voyages can be contrasted with more recent exploits like Columbus's first voyage to the Americas in A.D. 1492, and Thor Heyerdahl's voyage on the raft Kon-Tiki from Peru to the Polynesian islands in 1947, (over a distance of 4,300 miles).

In summing up the argument for the origin and inter-relationship of similar cultural elements and civilizations, it can be said that it amounts to a confrontation of the evolutionary explanation – which postulates the independent origin of similar cultures – and the historical and diffusionist explanation which proposes specific cultural origins and their spread. The thesis of progressive cultural evolution linked to biological evolution as indicated by Morgan, Bastian and Tylor, which has long been discarded and intellectually rejected even by established anthropology, nevertheless continues to cast its shadows and appears to be at the root of most of the theories which postulate the universality of independent cultural origins, whether they are of unilinear or of a multilinear evolutionary order. The latter process, as already mentioned, is explained by Glyn Daniel, as a 'super-organic, a cultural evolution in man's development'.[35] Its basic flaw is a wrong evaluation of natural selection, assuming that the latter was, and still is, perpetually active in the human species and that it causes cultural changes linked to a progressive organic evolution. The defect of such reasoning is the assumption that natural selection, especially if rigorous, is automatically causing evolution. In reality evolutionary stagnation is one of the most prominent features of the evolution of life. What we find is that a rigorous natural selection not only does not, as a rule, lead to evolutionary changes, but that it is necessary to keep species on existing levels of biological efficiency. Relaxation of natural selection on the other hand weakens the genetic stability of a species and leads to degeneration.

Take the example of our well-known parasitic companion, the common flea, which not only has accompanied the human species since its inception a million years ago, but pre-dates us by more than 35 million years. Fossil fleas found in the Baltic in pieces of amber 35 million years old are scarcely distinguishable from present species. Now, the life cycle of a flea (one generation) may last from eighteen days to several months. If we take one month as the average we can calculate that 35 million years represent 420 million flea generations. During all this time the flea must have passed through the most diverse climates and environmental conditions imaginable, undergoing a process of natural selection which can certainly not have been surpassed in rigour during any period of human existence. Yet all that this selection pressure accomplished was to keep the flea on the same evolutionary level. Compare these 420 millions of flea generations with the total of man's only one million years existence of c. 50,000 generations, and consider that natural selection pressure was considerably reduced during the last 15,000 generations (300,000 years); it becomes clear that there cannot have been any appreciable evolutionary change of a progressive nature in man during the latter period. It may be remarked that man's brain evolution nevertheless proceeded to some degree even during the latter relatively short period and that this is evidenced in the greater brain sizes of Neanderthal and Cro-Magnon man up to 35,000 years ago. Yet it is also evident that since then mankind has experienced an overall recession in average brain size.

This means that the evolutionary factor must be entirely excluded when we attempt to interpret cultural remains of a material as well as of a social (institutional) nature which occurred during the last 30,000 years, and certainly during the last 10,000 years. But once we exclude the evolutionary factor, the historical explanation for cultural origins, and the diffusionist explanation for cultural spread, remain the only rational alternative.

9

The Sovereign State

This chapter attempts to show that a correct interpretation of cultural history enables us to see also contemporary world political events in their true perspective. Among other things it destroys the myth that the sovereign state is an inevitable and permanent institution.

Thanks to the work of Professor Elliot Smith and J. W. Perry we are able to gain a more comprehensive picture of how the world's states arose. The thesis they advance is based on the contention that urban literate civilization originated in an area of the Near East which embraces Egypt and Sumer, and that as a consequence of it, the first state arose during the fourth millennium B.C. Subsequent cultures and civilizations of similar structure, including the later sovereign states, resulted from a gradual, and often sporadic, world-wide spread from this source. During this spread, geographical, historical and economic circumstances caused many changes. But the perpetuation of royal rule and the hierarchical structure of government and state have remained essentially the same throughout the ages.

The life of the early states revolved essentially round the personality of the king who derived his prerogative to rule by dynastic succession from the institution of 'divine kingship' (dating back to early Egypt/Sumer). The exact circumstances of its origins are still obscure. What is less obscure is that in the early states, the god-king was considered the benefactor of the country and his personal well-being was identified with that of the state.

The king was an absolute ruler and held power of life and death over his subjects. None of the latter, with the exception of personages of noble birth, took part in the running of the state. The position of these ancient rulers (and some modern ones) was essentially similar to that of the French Sun King. Louis XIV, whose saying 'I am the State' defines their status fairly clearly.

The ancient states of the world were not the result of racial affinities or of climatic or geographical factors but resulted from the dynastic processes of royal succession and intermarriage. It can also be shown that all over the world, with the exception of the supposed place of origin (Sumer/Egypt), most ruling groups derived their prerogative for rule from some pre-existing ruling group. It can be demonstrated that a considerable degree of dynastic continuity runs not only through the history of the entire ancient world, but also through later history.

For example, in early Britain all the royal genealogies of Kent, Wessex, East Anglia, Merovia, Lindsey, Deira, Bernicia, are descended from gods. Odin is said to have come to Denmark from Persia, where he was a member of the royal family. The Celts of central Europe, on the other hand, received their rulers and much of their culture from the Greeks. The Greeks in turn trace their ruling families and gods to Egypt. Danaus is supposed to have been the son of an Egyptian king and the nephew of the king of Phoenicia. The latter was himself a brother of the king of Egypt. Heracles, a hero of Greek mythology, came from Egypt where he was worshipped as one of the Egyptian deities.[1]

In modern history it is easy to show that most of the ruling families in Europe at one time or the other were interrelated. The British Royal House is related to the former rulers of Germany, the Hohenzollerns (which did not prevent them waging war against each other). The Hohenzollerns, in turn, had kinsmen on the thrones of Bulgaria, Greece and so on. The European ruling families formed a network that has no reference to nationality. A Hohenzollern (i.e. a Prussian) of possibly Italian origin ruled over Celts, Teutons and Mediterranean people. In the old Austrian Empire, peoples of Teutonic, Slav, Magyar, and Mediterranean extraction were under the rule of the Habsburgs. J. W. Perry concluded that 'the extent and life of any state in the past has been mainly dependent on the continuation of a particular ruling family, so that the study of the origin of war-like states in Europe really is the same as that of the origin of royalty'.[2]

The spread of early civilizations seems to have followed a certain definable pattern. Trading expeditions would set out to neighbouring countries, and from records which have been preserved we find that they were frequently led by men of royal birth. There was much search for gold, tin and copper and when such discoveries were made, the ancient explorers settled down to a period of construction, exploitation and consolidation. They would start to grow their own food, build houses, temples and palaces, and prepare burial places for their dead. The whole new system would, with local modifications, resemble that of their own mother civilization.

In most cases a royal prince would act as the governor of the new region and eventually would become the ruling king. The indigenous population, which meanwhile had been taught the arts and crafts of their conquerors, would carry on with their everyday lives, and would take little or no part in the direction of the state. Very often new settlements would extend their domain by pushing farther out.

It was a process very similar to the world-wide spread of European exploration and colonization which followed Columbus's discovery of America.

When the old civilizations and their dependencies increased in wealth and prosperity new motives made themselves felt which often erupted in internal power struggles. We know of the frequent dynastic quarrels for royal succession which broke out within ruling families of the ancient states. In many cases royal princes who failed in the struggle for the crown were forced to flee their home country. Occasionally they were able to establish themselves among peoples of a lower culture who lived on the fringes of the old civilizations. By taking advantage of their royal prerogative such fugitives and their retinue often imposed themselves on the indigenous populations and ruled over them as kings. They would not only teach them the elementary rules of civilization, but would also devote much time to training them in warfare. It would then frequently happen that these newly established kings, with the help of their new subordinates would attack their former homelands. If successful, they overthrew the established rulers, set up their own dynasties, and their sons would follow them in the royal succession. Out of such historical events, new states, headed by warrior aristocracies, came into being, breaking up and replacing the former archaic states.

Within the disintegrating old empires, there was no obstacle to prevent weaker monarchs from being overthrown by successful generals, as frequently happened in Roman history. Such usurpers replaced the king or regent, and some like Julius Caesar toppled the existing republican régime, and assumed the prerogative of royalty and would even have themselves deified. Occasionally, dynastic continuity became interrupted by republican régimes, as in the Roman republics, or in the Cromwellian period in England. In some cases republics would be followed by dictatorships, as frequently happened among the Greek city states and in Rome. There would also be cases where a monarch would rule with a parliament at his side (a constitutional monarchy). But history records that dynastic succession always showed great resilience and in many cases royal rule would make a comeback in some form. Even today when monarchic rule has been largely done away, there are still living pretenders to the thrones of Russia, France, China, Italy, Spain, Bulgaria, Greece, Germany,

Buganda and many others. While in exile or retirement these pretenders still hope and sometimes plot for an eventual return. There are also many royalties who up to now have managed to survive, either as absolute rulers or within the bounds of some constitutional arrangement. Recently, the Shah of Persia celebrated the 2,500th anniversary of the Persian empire founded by Cyrus.

If we accept Perry's thesis, it is clear that most of the sovereign states of the world resulted from dynastic struggles. Their beginnings date back to the early civilizations (Egypt and Sumer). On the other hand, many new sovereign states in Africa and elsewhere arose from a struggle for independence. Nevertheless in some cases they are the indirect result of the same process. The majority of these states have already come again under the rule of military dictatorships, and some of them may well relapse again into hereditary rule (e.g. Haiti).

One of the most remarkable features that accompanied the growth of civilization was the steady increase of organized warfare. According to Perry the civilizations of Central America, Mexico, Peru, Southern India and Java, all grew and flourished up to a certain point, as did those of Babylon, Assyria, Persia and that of the Hittites. Then something happened in every case, and the civilization decayed with greater or lesser speed. For example, in Rhodesia, India, the Malay Peninsula, Guyana, in the Pacific, and the Yenessei Region of Siberia, all the ancient mining activities of the original settlers, mostly living on a higher cultural level than the people who replaced them, seem to have come to a sudden end, and there is no trace whatever of a gradual degradation of culture. As a rule there is a sudden catastrophic event caused by a military invasion from without.

During the fourth century B.C. various Celtic speaking groups belonging to the Aryan-language family established their hegemony in central Europe. They succeeded in conquering Spain, but were stopped by the Romans. The Romans (themselves an Aryan-speaking people) fell later under the pressure of the Teutons and Goths, who with their incessant attacks almost obliterated the existing Mediterranean civilizations. What happened in this case was that Germanic princes received their military training by serving in the Roman legions. Later they used their military experience to conquer their former teachers. On the other side of the world, in Mexico, the Old Maya civilization was attacked and finally conquered by the warlike Aztecs, who came from the northern fringes of the country. Spanish conquerors in their turn destroyed the Aztec empire and established Spanish royal rule.

Perry rejects the notion that all the early war-like people

created their own cultures independently. He maintains that they originally derived their culture from more highly civilized peoples with whom they had previously come into contact. It can also be shown that most nomadic people were formerly agriculturalists. For example, in Asia the irrigators came first; the nomadic horse-riding people came later. When these war-like nomads arrived the mining activities of the original settlers came to a sudden end. The general principle was always the same. A great leader would arise in a small tribe, would conquer his neighbours, and then start out for greater conquests. Occasionally there appeared leaders like Ghengis Khan, Tamerlane, Attila, and Alexander the Great. They would conquer great areas of the world, but when they died their empires would rapidly disintegrate.

Many of the ancient civilizations never recovered from the onslaught of the barbarians, because the invaders frequently destroyed the water systems on which their flourishing settlements were based. For example, in A.D. 1258 Hulaga Khan, the grandson of Ghengis Khan, sacked Baghdad and his invading Mongol hordes destroyed the ancient irrigation systems of Mesopotamia. The harm has never been fully repaired, and even today the whole area is still semi-desert, dotted with the ruined remains of ancient cities.

The foregoing analysis, based on the work of Elliot Smith and Perry, shows that the birth of sovereign states was a fairly arbitrary process. In no case can it be shown that states came into being as the result of ethnic, racial or even economic considerations, unless, that is, the conquest of wealth and territory by kings is identifiable with the economic aspirations of their subjects. Instead, states arose from the originally dynastic struggles of ruling groups for power and domination, dating back five millennia. Furthermore, all through history the conception of the state as a sovereign entity has always been associated with war in which the ordinary citizen, metaphorically speaking, served as cannon fodder. It is, therefore, only natural that those historians and politicians who accept the sovereign state as a permanent historical phenomenon (as well as the less-informed masses) must believe that power politics and war are the natural state in human affairs. They do not seek the origin of war in the whims of aggressive ruling groups, but ascribe it to the inborn aggressive nature of man. As will be shown later, this is a mistaken view of man's nature.

Part Three

Natural Selection and Instinct

10

The Role of Instinct

The thesis of this book is that the motive of species preservation is the essence of all animal and human activities. In this process all biological characteristics have been developed by natural selection based on gene mutation. The basic living functions underlying this process are those of automatic reflex action, instinct and reason.

Instinctive urges are the primary forces that induce animals to act purposefully, but such action requires the aid of sensory perceptions. Sensory perceptions and their coordination form the basis of the mental process, and involve the capacity of reasoning. We can *thus* say that the functioning of a living organism depends on the joint action of instinct and reason while automatic reflex action serves as an indifferent auxiliary. It appears that under conditions of civilization the interrelation between instinct and reason has become disturbed in so far as the intellectual side of our nature has been overemphasized to an almost perverse degree, while the influence of our instinctive and emotional life has been neglected and even denied.

The concept of instinctive behaviour seems to have originated in antiquity with attempts to define a clear-cut difference between man and the animals. Human behaviour was said to be governed by reason, and the behaviour of animals to depend upon instinct. Possessing the unique power of reason, man was elevated above all other creatures, and, incidentally, his use of them for his own purposes was thus morally justified.

Christian theologians adopted this standpoint, taking the view that man was given the power of reason so that he could earn his own salvation. The lower animals were not accorded this power and their behaviour was explained as a product of divinely implanted instincts.

The relation between instinct and reason can be illustrated by the simple truth 'that a philosopher must also eat'; indeed he must eat first, before he can philosophize. Marx originally recognized this when he pointed out that the material needs of life are the prime motivation in human existence. Yet to satisfy the demands of a philosophical pet theory, namely that mankind is automatically progressing towards some ideal society, Marx created the abstruse edifice of historical and dialectical materialism. He thus reversed his former rational observations on the predominance of material factors, and subjected it to the mystique of abstract philosophy.

But Marx's philosophy is only one of the many philosophies that had their priorities wrong and set mental before instinctive motivation. By examining the life process it becomes apparent that our basic instincts are the primary forces necessary for the sustenance and continuation of life and that the role of reason is secondary. To satisfy the basic instincts, mental capacities serve as their aids and auxiliaries.

If we examine our cultural heritage, particularly the most treasured works of art and literature, we find that their prime motivation is instinctive. This is the case whether we consider the most lyric and tender feelings of love expressed by a poet like Heinrich Heine, or the dark passions of a Shakespearean play. Both are inspired by the powerful sex instinct. Also, the gentle expressions of motherly love, the feelings of friendship, the joys of companionship, as well as grief, hostility, revenge – all these feelings and more have their roots in our deep instinctive animal urges.

How the suppression of the hunger instinct can cause the extermination of a whole animal population is shown in the following example. As a result of natural selection acting through pesticides such disease-carrying, blood-drinking parasites as mosquitoes, tse-tse fly, and leeches have now become partly immune to conventional chemicals. Although more powerful insecticides could be developed, their ecological dangers exceed the harm that is caused by the disease carriers themselves. The Weizmann Institute of Biological Research in Israel has found an effective method to overcome both these difficulties. It has developed a chemical called azoester which is ecologically harmless but which affects insect pests in such a way that they lose their appetite. The insects will simply stop eating and die of starvation.

Likewise, if it were possible to suppress the sexual urge in a sexually reproducing animal species, the species would die out. Hunger and the sexual instinct need not always be jointly present to preserve the species, even in sexually propagating individuals. For example, neuter insects do not reproduce, and the biological activities of, let us say, worker bees are concentrated on keeping themselves alive, feeding the larvae, and maintaining the drones and queens which in their turn fulfil the task of procreation. In the mayfly the roles are reversed. This insect carries a very accurate clock and barometer. It only emerges at the vital moment of procreation under the right conditions of time, humidity, and temperature. It does not eat; indeed, it does not possess a mouth. After having carried out its reproductive function, the mayfly dies.

In most animals, particularly among the higher ones and in the human species, the instincts of hunger and sex reign supreme, however much the glamour of reason and of the so-called spiritual pursuits may be emphasized. Hunger and love are indeed the basic driving forces in the human species and there can be no healthy social life if this is not recognized. But just as the blood cannot pulse through the body without the pumping action of the heart, equally so the instincts are unable to function and fulfil their task unless aided by the activity of the senses which form the basis of all mental activities.

Unfortunately there is considerable confusion amongst different schools of psychology about the correct interpretation of 'instinct' and some adherents of the behaviourist school have even abolished the term 'instinct' altogether. Tinbergen, one of the pioneers of instinct research, speaks of a 'hierarchy of instinct behaviour', and Professor Lorenz speaks of the 'parliament of instincts'. In Freudian psychology instinct has become mixed up with such incomprehensible terms as the 'unconscious mind' and the 'id'.

As usual in biological activities, the proper clue to the instincts can be found in natural selection. A wolf cub is a good example of this. One of its first actions after birth will be the attempt to suck its mother's breast. Indeed, the instinct of hunger is essential, otherwise the cub would die. The same applies to the adult wolf. Once hunger (the urge to eat) arises in the wolf, the senses of smell, sight and sound are alerted to locate the prey. The senses of sight, smell and touch will guide the wolf towards it. By means of sight and touch the wolf will catch the prey and kill it. While eating the prey the wolf's senses of touch, smell and taste will discriminate between what is acceptable to its palate and what is repellent. The probably automatic reflex actions of biting, chewing, swallowing, and digesting will complete the act of eating. With the digestive process completed satisfaction will put the hunger instinct temporarily at rest until the requiremens of the organism again arouse hunger. This is, of course, a very simplified picture of how things work but it serves to show how the two 'musketeers', instinct and sensory perception, are dependent on each other, while reflex action, the third 'musketeer', acts as an essential, though dumb, auxiliary.

The apparent mystery of instincts arises from the fact that they cannot be described by means of a specific sensory experience. All we can say is that instincts manifest themselves in some deep inner urge, a need or desire, which is indiscriminate in so far as it depends for its guidance on the discriminative power of sensory perception. Furthermore, on analysis it is found that instincts are not memorized and can only be recalled by reconstructing in the brain the sensory circumstances under which an instinctive manifestation did occur. To remember an instinctive situation as, for example, a past experience of hunger, its recall must be triggered off by the memory of the sensory circumstances which surrounded it. Hence the technique of psychoanalysis uncovers the circumstances under which instinctive urges became inhibited or repressed. This is done by recalling from memory the original sensory situation, the circumstances under which it took place, the sounds uttered, the smells recorded, etc.

All this makes it obvious that the term 'instinct' must be restricted to a description of inborn urges and cannot be applied to describe specific activities or behaviour. Instincts manifest themselves in urges and desires which induce an animal to act in a purposive way to satisfy the instinct. Appetite is another expression to describe an instinctive desire, but the behaviour which it induces must be distinguished from the original urge of instinct.

The fact is duly recognized by Drever in his *Dictionary of Psychology* in which he defines instinct: 'in its original sense "animal impulse" (Gk., Latin, instinctus); hence a general term for natural or congenital impulse used in modern times for forms of behaviour rather than the underlying congenitive impulse (behaviour may be spoken of as instinctive, but not as 'an instinct'): it is more or less equivalent to innate or congenital "drive", or "need".'

There are different instincts and we have to discern between the primary instincts which are self-generating – those that can be aroused without exterior stimulation – and others that are not. Thus the hunger instinct may occur without any sensory awareness of food, or the male sexual urge without the sensory awareness of a female. But there are also instincts which are only aroused by some external stimulant. Such reactions may be partially or fully due to inborn stereotyped (automatic) responses of the nervous system which acts in response to a stimulant (releaser). On the other hand such self-generating instincts as hunger and sex can also be exteriorly stimulated. For example, once eating or sex has been experienced for the first time, the presence of food or of the female can promote and stimulate the instinctive urge and will arouse the appetite due to the recall of a previous experience.

In the male stickleback, fighting behaviour will be elicited when the fish sees merely the red belly of another male fish. But the response is so automatic that any similar red object, whether living or inanimate, will awaken the same response. Dr Lorenz has shown that a gosling develops a strong attachment towards the first moving object it sees after it is born and follows it. Normally it is the mother which is followed, but when Dr Lorenz substituted himself as the first moving object, the newborn gosling followed him instead, and a similar effect was produced by a stuffed model of a goose. Young nestlings will open their beaks as soon as they perceive a slight tap on their nest made by the parent bird and note its dark shape. But the use of a cardboard disc will produce a similar response.

A good example of the non-discriminating nature of the parental instinct has been given by A. Barnett. He described the nest of a meadow pipit with a young cuckoo in it. The young cuckoo had thrown the little nestlings out of the nest and they were squeaking outside it in full view and hearing of the parent birds. But the parents ignored their own young completely and fed only the cuckoo in the nest. In doing so, they were responding to the stimulus presented by the open beak of the cuckoo inside the nest.

There are many instincts in animals and some which *appear* to be instincts even though they may not be caused by instincts at all. Lorenz, who speaks of the 'great parliament of instincts', puts some of them into a special category calling them 'The Big Four', namely hunger, sexuality, flight, and aggression.[1] That hunger and sexuality are the most powerful instincts few will deny, but in the case of 'flight' and 'aggression' the situation is fraught with doubts. Flight in animals, for example, is either induced by the young imitating their elders, or by the endeavour to avoid pain and injury due to past experience. Some years ago a cinema newsreel was shown of a bird which had built a nest directly underneath the rails of a main railway line. Every now and then an express train would thunder across the rail but the birds would carry on feeding their young. If there was a flight instinct the parents would have been frightened away by the first noise of the train and could never have built a nest there.

Neither can it be proved that there exists a comprehensive intra-specific aggressive instinct in animals except that indications of truly aggressive behaviour, possibly instinct-inspired, are associated with sexual selection. But in the defence of a territory, and in the case of a mother animal protecting her young, aggression is elicited merely as a defensive stance. Should we therefore speak of an instinct or defence? The answer is that there are obviously different kinds of instincts at work, which, while inducing aggressive behaviour, cannot be attributed collectively to an aggressive instinct.

There are, no doubt, a great number of instinctive urges in animals which, besides the primary instincts of hunger and sexuality, play their part in the struggle for species preservation. Some of these induce animals to build nests, hoard food, care for the young, mark out territory, perform courtship dances, associate in groups, and many more. But in most cases it is not clear how the urge originates and at which point reflex response or reason intervenes. In this situation laboratory research cannot give a clear answer as many researchers themselves now admit. Hence, unless laboratory experiments can be supplemented by observations of animals in their natural habitats, the results must remain in doubt. In the majority of cases, it is difficult to decide whether certain behaviour patterns result from either instinct, reflex action or reasoning, or from a combination of some, or all, of them. Further, the fact that the terminology of instinct is often wrongly applied must leave the whole problem of inborn urges in animals unresolved, unless the definition of instinct advanced by Drever is unequivocally accepted.[2]

11

Human Aggression

In the human species, due to a sustained reduction in natural selection, many instincts found in animals have become diminished, or have altogether disappeared. In many cases human reasoning has replaced the need for blind, instinct-induced activity. One of these is the behaviour pattern termed aggression. Professor Lorenz in his admirable treatise *On Aggression* has described many conflict situations in animals which he compares with similar situations in man. However, close examination of human aggressive action reveals that there are few parallels. One animal trait which has been much misused in this respect is the defence of territory. Some authors, in particular Robert Ardrey, have tried to suggest that similar impulses, based on the defence of territory are active in the human species and are the principal cause of human warfare and aggression.

Before examining this assertion it is necessary to dwell a few moments on the term 'aggression' itself. This brings us back to Lorenz's book *On Aggression*, the original German title of which is *Das Sogenannte Böse*. Literally translated into English it means 'the so-called evil'. The German translation of the term 'aggression' would be 'Angriff', implying the use of unprovoked attack. If one reads Darwin or any other naturalist of the old school even up to the Second World War, one hardly comes across the term aggression. This term has only gained prominence since Adolf Hitler appeared on the scene and threatened various European nations with attack. Seen in such a context human aggression has little similarity to the conflict situations found amongst animals.

There are four principal conflict situations in animals which involve 'aggression' and which Lorenz compares with similar human situations.

(1) Carnivorous and predatory animals must kill other animals for food.

To do so some of them have evolved terrifying weapons. A wolf can rip the jugular vein of another, even much bigger animal, with a single bite; a crocodile can cut another animal in half with a snap of its jaws; a raven can peck out the eye of another animal with one single peck. Nevertheless, observers of predatory animals are quite unanimous in their view that such killings for food are generally done without much emotion or any show of ferociousness. This is mainly true for the killing of members of other animal species. A cat killing a mouse will do it without passion and so will a lion killing an antelope.

Lorenz draws some interesting deductions from this process. He points out that animals who can kill easily, like lions, wolves and others must have developed strong inhibitory impulses which prevented them from killing members of their own species. Otherwise, he believes, killings of their own kind would have led to species extinction. He further observes that animals like doves, hares or even chimpanzees, are not naturally equipped to kill their own kind with a single peck or bite or even injure each other seriously. Hence no natural selection pressure exists in inoffensively equipped species, including the human one, to produce inhibitions against inter-species killing.

In man, Lorenz points out, things took a different turn compared with the other inoffensively equipped species. The invention of artificial weapons upset the natural equilibrium that previously existed between the absence of natural weapons (thus automatically excluding inter-species killing) and lessened inhibitions against inter-species violence. In the absence of human inhibitions against inter-species killing man-made weapons became a means to kill fellow humans.

If we take Lorenz's reasoning to its logical conclusion we must say that from the time that man began to use deadly weapons, the uninhibited killing which this should have caused (in absence of inborn inhibitory checks) would have exterminated the human species long ago. Lorenz indicates that this did not happen because uninhibited killing between men was restrained by human reason.

In contrast to Lorenz's deductions it can be shown that the natural process must have been different. At first the increase of reasoning powers induced by natural selection led to the invention of tools and weapons (hunting weapons to kill for food and defend against predators). Secondly, the very invention of the tools and weapons made life easier, leading to a reduction in the intensity of natural selection. Decrease of natural selection, in turn, led to a decrease of inborn human inhibitions which, according to Lorenz, had previously served as safeguards against the killing of members of one's own species. So far we go with Lorenz. But the same decrease of natural selection must also have reduced man's inborn aggressive impulses, had they existed. Thus the balance between killer inhibitions (lowered) and aggressive impulses (also lowered) was maintained.

Such reasoning is confirmed by the observation that our primitive ancestors were rather peaceful and non-aggressive. Cave paintings dating back 30,000 years and showing hunting scenes but no human fighting, can be considered as a pointer to the fundamentally peaceful nature of man.

(2) Defence for self-protection is another conflict situation catalogued by Lorenz under aggression. It arises from the urge in animals to stay alive. It is the urge of the cornered rat to turn round and attack a much stronger aggressor. For the same reason a swarm of sparrows, otherwise defenceless, will surround a powerful sparrowhawk, mob it, and drive it away. Man, if cornered, will also act defiantly, and if there is a glimpse of hope, will sell his life as dearly as possible. On the same principle early man developed various weapons for self-defence, not only to be used individually but also collectively. However, action for self-protection has been and remains principally defensive and can therefore not be described as truly aggressive behaviour.

(3) One of the few truly inter-specific aggressive acts occurring in nature concerns the fight between male animals for a female for the purpose of copulation and procreation. In sexual selection the non-aggressive male will have little chance to leave any offspring. Lorenz rightly remarks that 'the strength of the father directly affects the welfare of the children in those species in which he plays an active part in their care and defence'. We assume that it must have also been the case amongst the family groups of our early ancestors. Thus, by sexual selection, strong family and social cohesion was maintained as long as natural conditions of life prevailed.

(4) Defence of territory. How does defence of territory work amongst animals?

Many animals (in a state of nature) when they intend to breed, mark out a territory so as to assure a food-supply for themselves, their spouses, and their offspring. Antelopes, for example, use glands with which they touch the ends of twigs. The smell clinging to the twigs serves as notice of ownership. Californian ground squirrels use the secretion from their dorsal glands for the same purpose. Civets and ocelots use their excreta, and many cat species use urine. Dung and urine are also used by hippopotami to define their territory. Outside their own territory animals seem to be at a disadvantage and it is generally agreed that a would-be invader of an already occupied territory will act more feebly and hesitantly than its adversary who is already in possession. What has also been observed is that the 'defence of territory' is an inter-species affair and that members of other species are rarely prevented from entering a territory already occupied by members of another species. But, if, for example, an animal should attempt to enter the territory already occupied by a member of its own species, a symbolic struggle will be enacted in which the contestants engage in a sham fight, without involving any real test of physical strength. This normally ends with the intruder beating a voluntary retreat.

We find, then, that 'the defence of territory' impulse, if closely examined, is not to attack, but to defend – as the phrase 'defence of territory' quite correctly suggests. It is a situation entirely different from human war-like aggression, which is as a rule, unprovoked, and where the invader is the first to use violence. Furthermore, while we can rightly speak in the case of an animal's 'defence of territory' as an instinct-inspired activity (which is hereditary), organized human war-like aggression is neither instinct-inspired nor is it hereditary. Therefore, animal examples of the 'defence of territory' are really not applicable to the human species.

It should be added here that under the artificial conditions of zoo life, in domesticated animals, and in the confines of aquaria, things work out differently. In such situations many unprovoked aggressive tendencies are displayed which cannot be considered as the natural expression of the animals involved. And the same holds good for the complicated conditions of human civilized life. In later chapters, some of the real causes of human conflict and war will be described.

Finally, we have to deal with the phenomenon of ritualistic behaviour which is both connected with the defence of territory and with sexual selection. It guarantees that male and female of the same species will recognize one another and it secures for the male the sexual compliance of the female. Such ritualistic behaviour consists in various stereotyped behaviour patterns which have evolved through countless generations by natural selection. Lorenz has given numerous remarkable examples of such behaviour patterns which Julian Huxley had previously compared with human rituals (a term which has now become common usage among animal observers). As it turns out, the comparison with human ritual is misleading.

In animals so-called ritualized actions, like the courtship behaviour of the great crested grebe or the collective courtship dance of the tailor fly (described by Brehm), are entirely instinct-inspired and hereditary. (Lorenz calls them phylogenetically determined.) Human rituals on the other hand 'which evolve in the course of human civilization, are not hereditarily fixed but are transmitted by tradition and must be learned afresh by every individual' (I use Lorenz's words.)

Curiously enough, Lorenz tends to disregard a similar distinction when he compares animal aggressive tendencies in which ritualistic behaviour is involved with human aggressive tendencies; because here also there is no valid parallel just as there is no such parallel between the behaviour of a bird building a nest and a man building a house.

With the extension of human brain capacities, natural selection pressure in the direction of stereotyped ritualistic behaviour patterns must have decreased. As a result we find that there are few traces of inheritable ritualistic behaviour even in primate society. Neither is there any specific ritualistic instinct in man relating to courtship or to the defence of territory. If there is, it has never been scientifically demonstrated.

Lorenz has done invaluable work in giving us a new insight into the complicated and often wonderful way in which animals reach their biological objective of species survival, but his many examples of aggression taken from the animal kingdom are really not applicable to the human species.

12

On the Nature of Violence and War

As Dr Lorenz and other observers of animals have found, violent aggressive behaviour, in the sense of unprovoked attack (with the exception of sexual aggression), is rare amongst animals in a natural state. There are examples to show that primitive races who live a natural life and are little influenced by civilization behave peacefully and non-aggressively. On the other hand, it is known that unnatural conditions of stress lead to violent behaviour both in man and animals.

L. Williams writes:

> The zoo colony in the past has amounted to little more than putting some fifty monkeys on a giant slab of concrete under the impressive title of 'Monkey Hill' or 'Monkey Island', with results that were often disastrous. As a result of overcrowding and unnatural stresses the London Zoo at one time lost sixteen Hamadryas females. They were literally torn to pieces by the males. The zoo colony provides no opportunity for developing real social or individual knowledge, such as we find in primate society in the wild. Bonds of affection cannot be formed on the basis of artificial security in a confined space. Degeneracy and tyranny take over, fighting and brutality become the determining factors and the means of establishing authority and rank and leadership falls to despots with no real sense of responsibility.[1]

Koestler, commenting on the same subject, refers to the observations of Goodall, who found little evidence of fighting among wild chimpanzees; nor did Hall find it in wild bands of those baboon species which Zuckermann had studied in the zoo. Emler and Schaffer did not observe the slightest trace of aggression in wild bands. Relations within different bands were so cordial that when two bands met they might bed down together overnight, and individuals would visit each other at any time. Koestler says that the early zoo findings made such a deep impression on field workers that they initially assumed that the species under study must be unusual. We can now see that they were wrong. All monkey and ape species are peaceful in the wild. Putting together the field and zoo reports, we now know that aggressiveness is not an innate feature of individuals, appearing in some primate species and not in others. All primitive species are peaceful in some conditions and violent in others. Violence is a property of societies exposed to stress.[2]

It seems that, despite appearances to the contrary, human nature is fundamentally peaceful and non-aggressive. Generally, man will not show violence unless attacked or conscripted and trained for warfare. Even then the majority of men will only kill with reluctance and revulsion. The civil war in Biafra is notorious for indifference to cruelty and suffering. Living in Nigeria at the time, I had many opportunities to speak to participants. Most of them had joined up because they had been under-employed or unemployed for long periods. None had joined because they felt any particular hatred against the Ibo tribe, or because they enjoyed fighting or killing.

This confirmed the author's previous experience during the Second World War when he served in the British infantry. His fellow soldiers were mainly conscripts and there was a good spirit of comradeship amongst them. At the time England was in a perilous situation. Yet the main topic of conversation among the soldiers was not about fighting and killing, but about girls. Most of them wanted the war to end quickly so that they could return to their sweethearts, friends and families as soon as possible. There can be little doubt that these sentiments are common to the soldiers of most armies, whatever their nationality or race.

Why is it, then, if men are of a generally peaceful disposition, that so many crimes, tortures and unspeakable massacres have been perpetrated in times of war and upheaval? Some of the reasons can be seen from the following personal experience. During the Second World War I was stationed in a detention camp where Germans suspected of war crimes were interrogated by British Intelligence officers. The prisoners were looked after and guarded by soldiers of the Grenadier Guards, the Scots Guards and the Coldstream Guards. Although some of them were a little rough in manners, they were otherwise friendly, amiable and normal people, their main concern being to get a week-end pass. The officers were cultured Englishmen. Yet on the instructions of the officers many of the prisoners were submitted to torture. The most common procedure was to duck them in cold baths and make them stand to attention for hours on end without allowing them to move. Another method was to starve them of food, which seemed the more effective as most of the prisoners had come from German camps, where they had already been on starvation rations.

While these tortures could be considered relatively mild compared with the unspeakable cruelties perpetrated by the Nazis, they were nevertheless carried out without the knowledge of the British public and with the compliance and participation of otherwise amiable soldiers. It happened that these incidents were occasionally discussed amongst the soldiers, and some of them at least expressed their personal disagreement with such methods, but all of them continued to obey the orders of their superiors.

No doubt the consciences of the soldiers were soothed by the knowledge that they were dealing with alleged war criminals who did not deserve normal human treatment; they therefore believed that they were fully justified in the interests of their country, and, alas, humanity, in subjecting the prisoners to brutal cross-examination. There was also the disciplinary code of the Guards, which would have punished any refusal to obey orders with long-term imprisonment. And, finally, public disclosure was prevented by the Official Secrets Act to which all soldiers were subject. There can be little doubt that it is similar circumstances that led to normally law-abiding citizens not only tolerating torture and massacre, but also participating in them. Thus we can explain the attitude of the German people towards the war crimes of the Nazis, the indifference of the Nigerian people during the Biafran war towards the starvation and suffering of the Ibo tribe, and the attitude of the so-called silent majority in the U.S.A. in face of the My Lai massacre.

Unfortunately, it is just this passivity of the vast majority of people of all races and nations and their inclination to avoid fights and conflicts, which time and time again has allowed ruthless, aggressive minorities, not only to impose control on fundamentally non-aggressive majorities, but to induce them to fight on their behalf. The many examples in history of unscrupulous minorities imposing their dictatorial rule on a non-aggressive majority by means of a small well-trained and willing fighting machine are evidence not of man's inborn aggressiveness but of his peaceful passivity.

ON THE NATURE OF WAR

Among the many prehistoric examples of cave paintings covering a period of at least 20,000 years, we find countless hunting scenes, but none depicting fights between human beings. It may be assumed that if there had been fights of men against men they would have been recorded. With the beginning of civilization (4th millennium B.C.), things begin to look quite different. In the early representations left to us by the Sumerians and the Egyptians we already find a predominance of human warfare while hunting scenes are rare.

Naturally, it would be against all visible evidence to describe man as an angel of peace. But considering him from the point of view of his family life, harmful aggressive tendencies do not appear to be his dominant characteristic. On the other hand there is little doubt that human aggression, as a neurotic reaction to the many unnatural conditions of civilized life, plays a prominent part in modern society.

Professor Elliott Smith and Dr J. W. Perry conducted much research into the conditions of life of primitive hunting and food-gathering family groups. They found that such people, although unable to escape entirely the influence of later civilizations, seemed to have retained most of the ingredients of social life which man must have possessed as his natural biological heritage. These hunter food-gatherers must not, of course, be confused with those savage agricultural or nomadic peoples whose acquaintance with archaic civilization saddled them with such savage relics as cannibalism, human sacrifice, head hunting and other barbarous practices.

Some authors have tried to ridicule the evidence of the fundamentally peaceful disposition of these people and decry the observations of Smith and Perry. One of the most outspoken has been Robert Ardrey who called the observations of Smith and Perry romantic fallacies. He says:

> It was Elliot Smith who provided the romantic fallacy with its most respectable scientific rationale. Like his contemporary, Dr W. J. Perry, he asserted that man's original nature could be studied if one observed those primitive people who by physical isolation or manner of life were most thoroughly cut off from all influence of civilization. The Eskimo, living on the frigid periphery of human existence, became a favourite object of such studies. The Yahgans of Tierra del Fuego became another such favourite, for the remoteness and hostility of the Yahgan environment could scarcely be rivalled anywhere on earth ...[3]

Ardrey concludes ...

> ... the shy, timid, amiable, non-aggressive nature of pristine man as has been revealed by the character of these people is a logical sequitur simply to a false premise. What has been revealed is nothing more than that people living where nobody else wants to live may quite possibly suffer from non-aggressive dispositions.[4]

In answer to Ardrey it must first of all be mentioned that many of the food-gathering people described by Smith and Perry live in quite accessible and congenial areas. They include the Punan of Borneo (Hose and MacDonald), the Veddahs of Ceylon (Seligman), the Chenchus in India (Heimandorf) and many others.

Professor Elliot Smith tells of peaceful as well as of violent behaviour found side by side amongst different Eskimo tribes, all racially identical and sharing the same environment. Professor Boas, the famous American anthropologist, writes of the Central Eskimo of Alaska: 'Real warfare or fights between settlements, I believe, have happened, but contests have always been confined to single families. The last instance of a feud which has come to my knowledge occurred about seventy years ago.' Another Eskimo tribe living further west is described as follows: 'Blood revenge is considered a sacred duty amongst them, and it is a common thing to find men who dare not visit certain villages because of a blood feud existing ...'

There are examples in New Guinea tribes of both peaceful and violent tribes living in the same inaccessible environment. One, the Arapesh, a mountain-dwelling people, appear to have outlawed competition, aggressiveness, and dominant behaviour, and to have replaced these with mutual trust and affection. They place little importance on age or sex differences, and their society comes perhaps as near as any to the ideals of a classless society. Even children's games are non-competitive. If fights arise they are immediately broken up by adults. Observers frequently refer to the Arapesh as 'one big happy family'. Another tribe, the Mundugunor, is cannibalistic, ruthless, violent and aggressive, and ready to fight at the slightest provocation. Their children are born into a hostile world, weaned early, treated to blows, and often killed by their mothers who show no signs of that hypocritical maternal instinct which is supposed to be a normal feature of femininity!

There are numerous other examples of peaceful food-gatherers, including members of races as different as the Australian, Negro, Negrito, and Mongol, and living under conditions ranging from the tropics to the Arctic. Professor Elliot Smith finds no reason for supposing that all these varied people have lost a culture that they once enjoyed. He considers it justifiable to assume that they are survivals of the state that was common to mankind before civilization about sixty centuries ago.

It is only since the Second World War that biological science has decisively rejected the Hitlerite thesis of inferior and superior races and of inborn aggressive and non-aggressive tendencies on a racial level. It has confirmed that mankind, as a species, has always been biologically fairly uniform and that there is no such thing as group violence based on genetic influences (although rare abnormal, genetically-influenced aggressive tendencies have occasionally been correlated with the presence of extra sex chromosomes).

Let us take the example of the Jews. While the Old Testament shows the Jews as a people continually battling with their neighbours, the picture changes radically as soon as we reach the Christian era. The Jews of the Middle Ages appear as one of the most dejected and inoffensive human groups. With hardly any trace of resistance they allowed themselves to be shut into ghettos. They were subject to countless pogroms and for hundreds of years became the *Prügelknabe* (whipping boy) of Europe. Under Hitler, millions of Jews were collected into concentration camps and later led into the gas chambers to be exterminated. And they took it with a surprising measure of meekness and absence of revolt. A few years later, some of the Jews who had survived the holocaust formed the state of Israel and took measures to defend themselves. The same Jews and their children who previously had submitted to all kinds of persecution, without a sign of counter-aggression, suddenly turned out to be the most successful warriors of modern times (considering their success in the Six-Day War and its aftermath). Can such a change in temperament and behaviour be attributed to a sudden genetic change which made the Jews aggressive? Of course not. Their change of attitude was entirely due to historical and educational factors which had nothing to do with any change in their biological make-up.

Perry refers to a constantly increasing education in violence and war as a direct result of the growth and spread of civilization.[8] For example, in early Sumer and Egypt, the king, when old and weak, together with all his servants and retainers, was ritually killed, for the supposed benefit of the community. At some later stage of history this practice was discontinued and other human victims, mostly war-captives, were substituted. In some communities, e.g. the Wakari of Northern Nigeria (the Junkans tribe), the slaying of the king is still in fashion.[9]

The need for human captives to serve as sacrificial victims is cited by Perry as one of the reasons for the spread of warfare. Head-hunting and cannibalism, as well as the taking of scalps by Red Indian tribes of America, fall into this category.[10]

Professor Elliot Smith and Dr Perry maintain that larger-scale warfare, which expresses itself principally in organized violent behaviour between states, is our heritage from the ancient archaic order and arose from the struggle of ruling classes for power and domination.

For example, if the early civilizations are examined it will be found that warfare occurs only sporadically. Regular large-scale fighting seems only to emerge from the Bronze Age onwards when conquerors like Sargon of Accad appeared on the scene (about

2,300 B.C.). Such warfare therefore coincides with the emergence and spread of war-like rulers, constantly on the lookout for new areas to conquer and dominate. As time passed, human history gradually became a continuous record of conflicts and wars.

13
Moral and Ethical Factors

Instinctive urges which still exercise a powerful influence on the human species are those concerned with parental care, the need for companionship and the desire for appreciation.

Some psychologists and sociologists have disputed the existence of a special social instinct, entirely fixed by genetical factors, and they may be correct. What is certain, however, is that what we call social impulses are activities which are initiated by both instinctive and by intellectual factors. Parental activities play a considerable part, and we can almost say that social feelings stretching beyond the bounds of the family group are merely the extension of those emotions that usually hold families together.

Modern social relations have indeed resulted from a direct continuation and extension of those bonds which were forged over the million or more years during which human family groups lived in a relatively natural state. This compares with the barely 6,000 years of corporate social life under conditions of urban literate civilization.

John Maynard Smith has pointed out that there is another evolutionary consequence of the fact that our ancestors have lived, probably for some millions of years, in small family groups.

> It means that natural selection can act not only by favouring one individual at the expense of another, but also one group at the expense of another. Characteristics could therefore be selected, not because they increased the fitness of the individual, but because they increased the chances of survival of the group to which he or she belonged.
>
> More important, groups whose members behaved in an ethical or altruistic manner, likely to assist others as well as themselves, would have a greater chance of survival. Therefore natural selection would favour the establishment of ethical practices.[1]

A study of primitive family groups, mainly food-gathering people, that survived into the last and even the present century, confirms that our basic social, moral, and ethical concepts are an integral part of human nature.

Morice von Overbergh, a missionary in the Philippines, has given a detailed account of the Negritos of Northern Luzon.

Two vices generally recognized as being common to many uncivilized people, are lying and stealing. It would seem almost against the nature of a Negrito to tell a lie, and one reason for this is perhaps his apparent inability to conceal his own thoughts. Whatever he knows he, like the *enfant terrible*, makes public. Stealing, if it occurs at all, is very rare. The fact that we left our belongings for several days where every passing Negrito could have taken from them whatever he liked, is sufficient proof of his honesty.

The pygmies of the Itori forest have no chiefs, and the basis of society is the family, taken in the wide sense of the term. The strangest thing about them – and this struck van den Bergh most forcibly – is that they are not devoid of any ethical and moral code. Indeed, the contrast of their ideas of morality with those of other African tribes is so great as to be astounding. Professor Elliot Smith writes:

It has for some time been a conviction with me that among most of the African tribes, especially those with which I have come into contact, there is almost a complete absence of morality ... I was, to say the least, surprised to find the Manbuti imbued with such high moral instincts. Stealing is so far foreign to their habits that the Wanyari Chiefs give them their goats and sheep to mind whenever a tribe lingers for any length of time in the same locality.

In his latest work on the Arunta of Central Australia, Sir Baldwin Spencer makes the following observations:

They usually wander about in small parties, consisting of one or two families, for example, two or more brothers with their wives and children ... As a rule the natives are kindly disposed to one another, that is, of course, within the limits of their own tribe and when two tribes come into contact with one another on the borderland of their respective territories, there the same amicable feelings are maintained between the members of the two. There is no such thing as one tribe being in a constant state of enmity with another, so far as the central tribes are concerned. Now and again, of course, fights do occur between the members of different local groups, who may or may not belong to the same or two different tribes.

Summing up his studies of primitive food-gathering people, Professor Elliot Smith concludes:

... the cheerful, happy frame of mind is really the normal condition in human beings. Manifestations of violence and anger are neither normal nor healthy. While our bodies are so fashioned as to enable us to fight in self-defence and to display anger and violence, such reactions are intended only for use in exceptional circumstances. A contented mind is 'the divine gift'. The society in which it exists is the healthy society. The conclusion suggested by the facts is that the type of behaviour in family groups is stable, happy, cheerful, lacking in violence. This may therefore be regarded as the standard of behaviour for human society.

Our complicated materialistic society has so corrupted human relations that it comes as a surprise to discover that man is basically a moral creature. Professor Elliot Smith has remarked that several leading scholars of his generation have written large treatises on the evolution of morals, which not only ignore the consideration that decent behaviour is an innate human quality, but express profound amazement that natural man should display moral qualities.

Superficially seen, such an attitude seems not without justification. Do we not every day witness acts of cruelty and inhuman behaviour? We read of parents sadistically beating their children, mothers abandoning their babies; we read of war, torture and the massacre of thousands of innocent people and are presented with the horrifying evidence in its most vivid details on television. We live in a world where the most extravagant luxury lives side by side with mass starvation and misery and where violent crime seems to be on the increase.

And yet, if we take a closer look at our next door neighbours, we find that the majority live in families. And when they do not, they endeavour to start one. Those who have children look after them quite selflessly; they feed them, house them and try to give them the best possible education. And when they are sick they fret over them and do everything they can to make them well again. In short, the vast majority of parents feel the need to see their children happy. This is obviously done for selfless reasons, unless we call the satisfaction which someone derives from seeing someone beloved being happy, selfishness. Thus the overwhelming majority of men and women, whatever their race or nationality, work and suffer quite selflessly for their family. True, from a man's point of view, the primary motivation to take a wife is the satisfaction of his sexual urge. But once children are born and live in close relation with father and mother, parental feelings of love and affection will exert themselves; and this will happen although it may adversely affect the individual's comfort and the material well-being of the parents themselves.

But in an industrial society the true expression of human nature remains largely hidden behind a veil of indifference which everybody has to create in order to shield himself and his feelings from the cruel distortions which modern life and its artificial values impose. This is a situation unprovided for in our genetic make-up, and can therefore only be resolved by consciously designed social changes, which are adapted to meet our biological needs.

To sum up we can say this about social cohesion: The strongest ties that bring and hold individuals together are the instinctive urges for procreation and the family. The emotional cement of such unity is mutual love and devotion. From a biological point of view, these are the basic sentiments which exist in people all over the world, in spite of adverse signs. If people could be convinced that such sentiments can be extended beyond the narrow confines of tribe, race, and nationality to embrace the whole human family, then world unity could become a reality, not only in historical but also in political terms.

Part Four

Natural Selection and Reason

14
The Role of Reason

The satisfaction of instincts, which we have seen are primary urges, depends on the aid of sensory perception. When these activities of the senses are analyzed, their coordination represents the phenomenon which we call the 'mental process', consisting of our powers of thinking or reasoning.

How do sense organs function? It can be presumed that an original primitive, discriminative sensitivity towards nourishment must have been the precursor of the sense of taste. In the same relationship stands light to vision, sound to hearing, pressure and other skin irritants to touch, and odours to smell. There exist also senses, like the echo sense in bats, or a time sense in birds and fish, which are orientated towards the sun and are outside the scope of normal human sensory perception.

The feeling of pain is of extraordinary importance for an individual's survival. Without it individuals would recklessly plunge into dangers, injuries would not be noticed, and even the loss of whole limbs would go unnoticed. It is known that lepers, having lost the sense of touch in their hands, are not sensitive to injuries or burns, and as a consequence frequently show hands, the fingers of which are only short stumps.

The senses of taste and smell are also influenced by pleasurable or distasteful reactions. The sexual act, for example, is accompanied by one of the most powerfully pleasurable sensations. On the other hand, the senses of vision and hearing are not generally guided by pleasure or displeasure, but human experience shows that visual and auditory tastes can be acquired and cultivated.

How is the mental process accomplished? To be of any use, senses must be able to discriminate between perceptions of different intensity. But this is only possible if sensory perceptions are separated in time. The bridging of this time-gap requires the retention and storage of sense impressions, or, in other words, the possession of a memory. Further, the comparison and evaluation of differing sense impressions, linked as they are by memory, amounts to the capacity of thinking or reasoning. It can therefore be asserted that even the most primitive form of sensory perception presupposes the possession of a memory and the capacity of thinking. Sensory perception existing without memory would be of no practical use whatsoever to any kind of living organism.

Some of the most primitive organisms, such as bacteria, viruses, and protozoa, exhibit traces of sensitivity, going beyond automatic reflex action. All these primitive life-forms must, therefore, possess memory, and though it may sound far-fetched, they must also be capable of thought. Such lowly organisms possess, of course, nothing like a brain. But they must inevitably possess some primitive nerve centre as the brain's forerunner, capable of associating and utilizing sense impressions. Thinking and memory do not necessarily require a complicated organ like the brain.

Writers such as E. Hering believe that all organisms have memory, in as much as the action of a stimulus on protoplasm always leaves traces which influence further response. In this sense, the term learning may be extended to the simplest types of behaviour changes through stimulation, as in plants and protozoans. In the words of the 1961 edition of *Encyclopaedia Britannica*: 'If mind be granted to the higher vertebrates, the argument from the observed continuity in mode of behaviour indicates that it must be granted to the lower vertebrates and indeed to all animals including the protozoa. No distinction, moreover, can be made between the protozoa and the protophyta.'

Even such a lowly organism as a virus, then, may be endowed with the rudiments of thinking. Now, viruses are generally so small that they pass the finest bacterial filters (bacteria themselves were once thought to be the smallest entities capable of independent life), but some of them carry in their chromosomes as many as six or seven thousand genetic determinants. The question is, why should such an utterly tiny organism as a virus carry so many thousands of genetic determinants?[1] The answer can only be that natural selection, considering the intense competition which must prevail amongst rapidly reproducing viruses, would not have tolerated the accumulation of so many thousands of genes unless they had been beneficial for survival. We must, therefore, assume that the thousands of genes of a virus control the thousands of different living processes that are necessary to keep it alive, amongst them in all probability genes controlling sensibilities, memory, and other aspects of mental action.

It should eventually become possible to locate more and more genes in the germ cells of animals and determine their functions. We may even predict that it will eventually be found that certain of the genetic determinants controlling nervous reactions, sensitivity, thought processes, and memory, in animals as well as in man, are already present in viruses and in bacteria.

It may even be discovered that the genetic system of a virus is more subtly balanced than that of man. The reason is that their rapid increase subjects them to an immeasurably more intense process of natural selection than the slower-breeding animals of greater complexity.

For example, one human generation lasts about twenty years while a virus generation lasts only about twenty minutes. This means that one human generation is paralleled by about 500,000 virus generations.

Consequently, the human species would have to exist for 10 million years to be exposed to the same hazards of gene mutation and natural selection that viruses experience within a time space of only twenty years. But true man-like species are hardly a million years old.

To return to our main discussion, it can be shown that the distinction between sensory reflex action is that the latter proceeds without memory. Thus reflex actions, like the heartbeat, the action of cell growth, the activity of breathing, of digestion, the action of swallowing food etc. are automatic and are not memorized. The reason is that reflex activists have no discriminative functions and therefore need no memory.

Sensory perceptions, on the other hand, have discriminatory functions and depend on memory. As a consequence, we find that a reflex action can short-circuit the brain, and, like instinct, is not involved in any thought process. On the other hand, even in unicellular life as well as in multicellular life (up to a certain stage before the possession of a brain) thought processes must take place. However one looks at the process of thinking or reasoning, it is basically nothing more than the coordination and association of differing sensations, firstly, by their quality and, secondly, in their intensity. Sense organs without discriminatory capacities would be an absurdity.

We see, then, that thinking, reasoning and intellectual activity all express the same capacity of living organisms to compare and associate sense impressions of different kinds and of varying intensity. The primary objective of this combined action is the gathering of experience or learning in order to anticipate the vagaries of future situations. The ultimate aim is, of course, the maintenance of life and the preservation of the species.

Darwin has described an experiment with a pike which was put in an aquarium stocked with small fish, but was separated from them by a plate of glass. Driven by the urge to eat, the pike tried to catch the fish. But in so doing, he repeatedly ran his head against the dividing glass. Very often he was completely stunned by the impact, so that he eventually gave up. When the dividing glass was later removed, the pike, who could now have the fish, ignored them. However, when a different type of fish was put into the aquarium the pike did not hesitate; he caught and devoured them.[2]

This example shows the presence of a complete thought process, where past experience in association with memories of pain served as a guide for future action, while the original cause of the whole chain of activities arose from the hunger instinct.

Other conspicuous examples of animal intelligence are certain ant species which use the plantjuice-sucking aphids as milk cows. These ants regularly tap each aphid with their antennae, forcing it to exude drops of amber-coloured milk (honey-dew) from the horns of the abdomen. During the summer the aphids feed in the open on all kinds of plants and trees, while the ants herd them like cattle. When the aphids are moved to fresh pastures, they are not driven but are carried, each one bodily, in the ant-herdsmen's jaws. Before winter comes, they are taken below ground, where they are fed on the stems of grasses or other roots which grow in or near the ants' nests. In this way the ants are provided with winter 'milk' just as man keeps his cows over the winter in a warm stable and feeds them.[3]

Koestler mentions the example of the female of the wasp, *Eumenes conica*, which builds clusters of clay cells or pots, deposits an egg in each, provisions it with caterpillars for food, then closes the cells with clay lids: 'If now an artificial hole is made in a cell, the wasp will first stuff the caterpillars which have fallen out back through the hole, then mend the hole with a pellet of clay – operations which are quite different from the normal building routines.'[4]

Koestler writes, that in bees ...

> ... if one of the specialized age-groups is artificially eliminated from the colony, a kind of collective super-flexibility manifests itself in the hive: other age-groups deputize for the vanished group and thus save the super-organism. When, for instance, all pollen-and-honey foragers are taken away, usually bees of twenty days, the younger bees of scarcely six days old, who normally feed the larvae, fly out and become foragers. If all building workers are taken away, their task is taken on by older bees who had already been builders before, but who had gone on to the stage of forager. To this end they not only change their behaviour, but also regenerate the wax-glands. The mechanisms of these regulations are not known.[5]

Carthy has cited evidence showing that some birds and mammals can recognize numbers. Beyond this they can match numbers and can remember them, and they will utilize that memory in future situations. For example, ravens, squirrels and parrots have been taught to take food from one particular bowl amongst a group of bowls distinguished by different numbers of black spots.[6]

Thus it can be seen that, while the mental capacities of animals differ from those of man in degree, they are basically of the same potentiality. But in the absence of anything resembling human articulate speech, their usefulness has remained relatively limited.

COMPARATIVE BRAIN FUNCTIONS OF ANIMAL AND MAN

S. H. Romer compares the brain functions of frog and man as follows:

> In their nervous tissues the spinal cord, the peripheral nerves, and the automatic nervous system, the frog and man are essentially similar. Apart from the brain, the only noteworthy difference is that in the frog the body is so shortened that there are only ten pairs of spinal nerves, while man has three times that number. Only in its most anterior portion is the frog brain persistently primitive and differs greatly from that of man. Furthermore, the ten cranial nerves in the frog compare in major features with the first ten found in man. Man possesses a well-developed twelfth nerve, which is absent in the existing amphibians, but is found in reptiles, birds, and other animals. Its absence in the frog has often been considered as a primitive character. Evidence shows, however, that it was already present in primitive amphibians; so that this and other features of the frog must be considered as degenerate.[7]

It is generally held that the most conspicuous difference between the human brain and that of an ape and other mammals is its size. However, the giant whale and even the smaller elephant both possess bigger brains than man. But in man the brain weight is about one forty-sixth of the body weight, while in an elephant it is only about one five-hundred and sixteenth and in a large whale about one eight-thousandth. In contrast, the twenty-foot-long stegosaurus possessed a tiny head with a brain the size of that of a kitten.

What do such examples suggest? First, if the relation between body weight and brain weight is thought to be of decisive significance, then the brain of some insects must be vastly superior to the human brain. Secondly, a relatively smaller brain within a large animal is not necessarily inferior to a relatively bigger brain in a smaller animal. If it were practicable to implant the smaller human brain into a whale it can be assumed that it would be quite adequate to direct and serve the much bigger animal just as a relatively small radio transmitter can serve equally well for a small boat and for an ocean liner.

One should further assume that increased memory in connection with a greater vividness and a greater range of sense impressions could eventually lead to an increase of reasoning powers. But this is not necessarily the case as the following example shows. It concerns a Mr S., the owner of a phenomenal memory, whose case was studied over a period of years by Professor Luria, a famous Soviet psychologist. Luria's studies were published under the title *The Mind of a Mnemonist*. The principal facts about S. were these. His memory was so perfect that sixteen years after the event he could still reproduce long and complicated sentences of books which he had read. Professor Luria examined S. regularly over thirty years and found that the achievements of his memory depended on a remarkably vivid sensitivity or synaesthesia. When S. heard a sound, it immediately created a bright experience of lights and colour and even of taste and touch and, similarly, other stimuli were experienced by all, and not just one, of his senses. Passing a fence on a walk, he remarked that he knew he would remember it because it had such a salty taste and felt so rough or it had such a sharp piercing sound. The greatest disadvantage for S. was the difficulty of forgetting. His profession probably didn't help him (he showed his memory feats on the stage) and the pressure on his reasoning powers became so great that he had continuously to invent various methods in order to forget. But S. was also handicapped in the understanding of poetry because the images it produced were so strong that he accepted them literally. At the same time abstract thinking was almost painful for him and his reasoning was very child-like and naïve. But he had such a powerful imagination that he could increase his heart-beat by merely seeing himself running to catch a train. This example shows that a superior memory and a more vivid imagination are not necessarily synonymous with more rational thinking or greater intelligence. On the contrary it shows that the capacity to forget is just as important an attribute of the brain as is memory.

Mental power in animals has been evolved to provide foresight to deal with new situations as they arise continually in life. The basic objective is the accomplishment of the vital, instinct-inspired pursuits of eating, self-protection and procreation. As the future is unknown, any guess about it can only rest on past experience. The prediction of future situations must, therefore, always contain an element of uncertainty. Fundamentally, mental prediction is a guessing game, and, as such, a shorter-term forecast reduces the margin of error.

Once a certain degree of intelligence had been gained, it is doubtful whether a further growth in mental powers could increase the degree of foresight. If it were not so, many animals would probably have evolved super-brains long ago. It appears therefore that mankind has reached some kind of mental Rubicon.

It has been shown that in the early evolution of the human species many anatomical weaknesses were compensated by a bigger brain. But it is probably certain that its maximum benefit was reached with the establishment of primitive family groups. When human groups developed later into larger units, collective planning or long-term prediction went increasingly astray, becoming diverted from the first biological objective – that of species preservation.

In contrast to humans, we can say that animals, having less mental scope, have a smaller range of foresight than man. In a given situation their mental choices are also fewer. Animals will, therefore, as a rule, act more spontaneously and decisions will be carried out with little hesitation. Thus animal actions may often appear instinct-inspired though they are entirely guided by rational considerations. As natural history shows, animals got on quite well with a simpler, smaller brain. The proof is that some of the lowest living forms have survived for several hundreds of millions of years without possessing any brain at all and they are still going strong. Man, in contrast, though possessing a relatively huge brain, is hardly a million years old and his survival is already seriously threatened, and this significantly enough is a consequence of his super-brain.

Measured by long-term survival considerations, the human brain must therefore be judged to be, at best, of equal but not of superior quality to the brain of animals. Also, while it is admitted that the range of human intellectual activities is vastly greater than that of any animal, its very complexity must on balance be considered to have its drawbacks. It is already evident that man's possession of a super-brain, in spite of all its attendant advantages, has pushed the human species into a number of biologically harmful situations. Although it has brought about modern civilization, the consequent side-effects are presenting us with an environmental crisis which is not directly related to genetically induced inheritance. As a consequence mankind finds itself in the position of the sorcerer's apprentice: the childish optimist (now on a global scale) who unleashed forces which first appeared highly beneficial, but later turned out to be destructive and beyond control.

15
The Syndrome of the Sorcerer's Apprentice

The first damage caused by the possession of our super-brain was a result of the creation of cultural aids, which, by reducing the force of natural selection, led to the spread of genetically induced hereditary defects. This process began at least half a million years ago when men begun to use fire and clothing.

Man's blundering on a larger scale was precipitated by the so-called agricultural revolution. This event has been hailed as one of the most magnificent achievements of the human species. By growing more food than the growers themselves could consume, great numbers of individuals were liberated to occupy themselves with intellectual pursuits. They succeeded in developing the innumerable arts and splendours of civilization which eventually led to modern science and technology.

Ever since the inception of agriculture countless millions of people have been working the land with bent backs, hoeing, tilling and weeding the fields. But man, long before agriculture began, had been elevated by evolutionary means to live a two-legged upright existence, well suited to do a lot of running, jumping and even climbing. Thus he was not orthopaedically well adapted either for a life requiring forward-bending back-breaking work or for an existence which includes long periods of a sitting posture. The result in the former case has been the misuse of the spinal column. Professor Cyriax, a renowned orthopaedic authority, has treated this problem in his book *The Slipped Disc*. He has pointed out that any weight-bearing bending activity of the spine, if unsupported, is bound to injure the delicately balanced spinal column, by damaging the elastic cushioning discs which keep the vertebrae apart. As a result the discs will break and protrude onto the spinal nerves causing the excruciating and often life-long back pains of sciatica and related complaints. Further, a prolonged sitting posture is bound to shorten the stomach muscles, immobilize the joints, and lead to their degeneration. This leads often to painful and sometimes life-long debilities going under the general terms of rheumatism and arthritis, a condition affecting millions of people.

Fortunately for all these sufferers there is no such thing as Lamarckian inheritance. So that while a person may be a life-long sufferer of these debilities his children will again be born healthy, and at least be unaffected by those complaints which his parents have acquired during their lifetime.

Another consequence of the agricultural revolution unsuspected by its perpetrators relates to the staple diet of cereals and tubers. Too much reliance on this food alone has been the cause of chronic protein deficiencies for whole populations. It affects hundreds of millions of Asians whose main diet is rice and who cannot afford the supplementary protein foods of fish, eggs, and meat. It also affects the millions of Africans who live principally on the starchy diet of yam and cassava. Thus hundreds of millions of people suffer from deficiency diseases and malnutrition.

Another consequence of the agricultural revolution has been the steady increase of populations. Coupled with an improvement of sanitation and under the influence of preventive medicine this had led to the present population explosion.

All through the hundreds of thousands of years of the Stone Age, the human population was probably stabilized at a world figure of between 2 and 10 million. Within the last 10,000 years the world population has increased to over 3,000 millions and the estimate for the end of the century is 7,000 millions.

In 1859, Darwin wrote that in about a thousand years there would hardly be any standing room on Earth if the population growth continued at the same pace.[1] Now it looks as if Darwin's estimate was on the conservative side. According to some calculations, a child born today, living on into his seventies, would know a world of 15,000 million people. Standing room on this planet would be exhausted in less than 500 years.

To deal with the present population increase, the United Nations Economic and Social Council proposed in 1970 that the developing nations should build at least ten units of housing per 1,000 people annually. At the same time it was pointed out that in many countries only two units per thousand are at present constructed.

Nevertheless, populations are relentlessly exploding in what the UN report terms the 'unexploding economies'. It is predicted that in the next decade, eighteen Latin American cities will expand to one million or more inhabitants, whether the nations affected are prepared for this human flood-wave or not. Bombay and Calcutta might swell to 20 million or even 30 million residents by the end of the century. All these predictions are made after allowing for existing measures to reduce population increases, like contraception and sterilization.

We often speak with derision of our primitive hunter-gathering forefathers who had no houses and lived in the forests with only the sky as a roof, or of others who lived for many thousands of years in the shelter of caves. Yet in contemporary Calcutta the nocturnal streets are littered with hundreds of thousands of sleeping bodies, exposed to the contaminated air of the big city and without any visible means of sustenance.

The first and most immediate requirement of a population increase is the need to increase the food supply. Other requirements are more living space in the form of housing and open spaces for recreation. Then there follows the need for the basic amenities of life which can only be satisfied by industrialization. This, in its turn, necessitates educational facilities to supply managerial skills and a large working force to deal with the more menial tasks. But it also needs a large power supply to turn the wheels.

It can be seen that some requirements, namely the need for more agricultural land for food growing and the need for more living space and for industrial land, are competing with each other for the restricted space available. Additionally, large-scale industrial expansion is depleting the earth's natural resources and threatening our small planet with pollution.

Thus, while nutrition experts are working on the problem of how to feed the steadily increasing population of the world, population increases are rapidly devouring the cultivable area. It is said that in the United States of America erosion is destructively active on 775 million acres. About 100 million acres of arable land, much of it once the best land in the United States of America, is said to be already exhausted and beyond restoration. It takes nature from 300–1,000 years or more to bring back a single inch of topsoil. Every other continent has suffered erosion and it is believed that a total of nearly 2,000 million acres is now affected.

Of considerable help in increasing the food supply of the world, besides the development of higher-yielding strains of plants and cereals, has been the extensive use of pesticides and artificial fertilizers. Here particular mention must be made of the pesticide D.D.T. because its use throws an illuminating light on the lack of intellectual foresight of modern man in the field of science. Thus, when in 1948 the Swiss chemist Paul Müller received the Nobel Prize for his discovery of D.D.T., he was considered a benefactor of mankind. Twenty years later, the benefit of this and other pesticides has become a doubtful blessing. D.D.T. and the substances derived from it are now among the most widely distributed synthetic chemicals. They have been detached far away from the sites of application in soils, air and rainwater and in animals all over the world, including Antarctica. D.D.T. has recently been found in the Bermuda petrel, and may be the cause of the decline in their numbers. The petrel takes its food exclusively from the ocean, which means that a lengthy ocean food chain, presumably including plankton, is contaminated with D.D.T.

Pesticides such as D.D.T., parathion, aldrin and dieldrin are both ally and enemy to man. The chemicals annihilate predators: the aphids that plague rose-fanciers, disease-bearing mosquitoes, beetles that spread Dutch-elm disease, insects that devour crops. As a farmer's helper, pesticides increase crop yield, hence profits. But poison is blind. Loosened annually by the ton from planes, boats, trucks, tractors and hand-spray cans, it cannot isolate its target. To counteract their pernicious effects, several governments, among them that of Sweden, have prohibited the use of D.D.T. and others have severely restricted its use, although in some cases D.D.T. has almost completely wiped out malaria.

Industrialization, the boon of the last century, is now looked at by conservationists in the highly industrialized countries with considerable misgivings. But in the underdeveloped world it has not yet lost its glamorous appeal and is proclaimed as a certain panacea. It is considered to be a kind of magic wand which will cure the ills of under-development and backwardness, and provide work and incomes for all.

Yet the results of an intense and unrestrained industrialization are clearly demonstrable in a country like the United States where its dangers are now expounded by countless articles and television commentaries. This is how *Time* magazine describes the industrial plight:

> The result of massive production is massive filth. Every year, Americans junk several million cars, 100 million tyres, 20 million tons of paper, 28 billion bottles and 48 billion cans. Just to collect the garbage costs 2·8 billion dollars a year. The U.S. also produces almost 50 per cent of the world's industrial pollution. Every year, U.S. plants discard 165 million tons of smoke and fumes into the air.
>
> Modern technology is already burdening nature with tens of thousands of synthetic substances, many of which almost totally resist decay – thus poisoning both man and his fellow creatures. The burden includes smog fumes, aluminium cans that do not rust, inorganic plastics that may last for decades, floating oil that can change the thermal reflectivity of oceans, and radioactive wastes whose toxicity lingers for literally hundreds of years. The Earth has its own waste-disposal systems, but it has limits. The winds that ventilate the Earth are only six miles high; toxic garbage can kill the tiny organisms that normally clean rivers.[2]

But the rest of the world is equally affected. It has been pointed out that a million tons of oil are dumped into the Mediterranean every year, while the total poured into the world's oceans amounts to 100 million tons. When the Norwegian author-explorer Thor Heyerdahl sailed across half the Pacific on a balsawood raft twenty-two years ago, he recalls, 'We on Kon Tiki were thrilled by the beauty and purity of the ocean.' During his first attempt (1969) to sail from Africa to Central America, in a boat made of papyrus reeds, which he was forced to abandon 600 miles from his goal, Heyerdahl's old thrill was replaced by shock. What most appalled him were sheets of 'pelagic particles'. At first he assumed that his craft was in the wake of an oil tanker that had just cleaned its tanks. But on five occasions he ran into the same substances covering the water so thickly that it was unpleasant to dip even a toothbrush into the sea. Occasionally, the water was too dirty to wash the dishes in.

Population increase, with its twin increased industrialization, also creates an insatiable appetite for power supply. The success of the communist revolution, said Lenin, shortly after the Bolshevik take-over in Russia, is synonymous with the Soviets plus electrification. Now more than fifty years later, while the Soviet state has become consolidated, the need for electricity and other sources of power has remained just as great, or is even greater than it was in Lenin's time.

It has been calculated that in the United States the nation's need for electric power is doubling every decade. Massive use of cooling water to condense steam is an integral part of the generating process. Almost half the water used in the United States is for industrial cooling. By the year 2,000, the need will equal two-thirds of the nation's natural daily water run off of 1,200 billion gallons.

While the problem of radioactive seepage from nuclear power stations raises a lot of disquieting questions, the problem of radioactive waste-disposal from plants which provide the raw material for atomic and hydrogen bombs is even more serious. This is all the more alarming, as the military establishments in all countries where such war preparations are taking place are cloaked under a veil of secrecy, concealing all kinds of undisclosed dangers, which may or may not show their effects later.

According to a report in the *Observer* (London, March 1970):

> ... the bulk of the Atomic Energy Commission's radioactive waste lies buried in 140 huge steel and concrete tanks at Hanford Reservation in the State of Washington. It is estimated that about 50 million gallons of radioactive waste are stored there, continuously boiling from self-generated heat. An artificial cooling system has to be continuously operated to prevent the boiling fluid from bursting the tanks. If the tanks were ruptured, the wastes could find their way through the

ground-water to the Columbia River and out into the Pacific Ocean.

But this is only one isolated example. There are many more nuclear waste-disposal grounds in America alone (huge quantities have apparently been dumped into the sea). In addition, there must be a number of such dumping grounds throughout the Soviet Union, China and France, and perhaps in several other countries, which at the present time openly or clandestinely experiment with nuclear material.

The danger of playing with nuclear energy is also visible from the following isolated example. When United States B 52 bombers carrying H-bombs crashed in February 1966 in Palomares, in Spain, and in January 1968 in Thule, Greenland, each plane was carrying four H-bombs. In Palomares, one fell into the sea, and was recovered only after an 80-day search; two of those that fell on the land fragmented, spilling enriched uranium and plutonium. The clearing-up exercise cost 30 million dollars and involved shipping 1,000 tons of contaminated soil to North Carolina.

At the time of writing the United States and Soviet Russia are involved in a nuclear arms race which has been going on for over twenty years. Several years ago, China and France joined them.

While American strategists and scientists, as a consequence, talk about megadeaths (i.e. units of a million deaths in the event of a nuclear war) Russian and Chinese leaders have not been less outspoken. For example, Mao Tse-tung is reported as having stated at a meeting with Russian leaders in Moscow as far back as 1957 that in a war with atom and hydrogen bombs perhaps one half or three-quarters of the world's population may be wiped out. At the same meeting a former Chinese Defence Minister, Peng The-Huai, expressed the view that he would be quite satisfied if 200 million of the then 600 million Chinese survived. He said: 'Because, the losses of the Imperialists would be proportionally bigger, China would still win the war.'[3]

The effects of the more recent examples of nuclear hardware can be guessed from a report of the Atomic Energy Commission, details of which were disclosed by Ralph Lapp, a physicist formerly in the pay of the American Defense Department. In an excerpt published by *Time* magazine in 1968 Lapp speaks of experiments with rabbits relating to trials with anti-ballistic-missiles (ABMs). Exposed rabbits suffered retinal burns at slant distances of up to 345 miles from the blast. Furthermore, it was calculated that there would be some danger to human beings at distances greater than 200 miles under similar circumstances. Lower altitude explosions in the atmosphere would be even more disastrous. Lapp calculated that the detonation of a ten-megaton intercontinental-ballistic missile (ICBM) by an intercepting 'Spring' anti-ballistic missile at an altitude of 50,000 feet would produce second-degree skin burns in people over an area as large as 2,000 square miles and cause dry paper to ignite over an area of 3,000 square miles.

It should be noted that the above experiments merely relate to effects caused by one missile of the ten-megaton range. Present stock-piles suggest a much greater holocaust and can be placed in the 'overkill' bracket, meaning that present stocks contain the capacity to kill the populations of the U.S.A. and Soviet Russia several times over. According to figures given to the Richardson Institute (the London-based conflict research organization) by control specialist Dr Frank Barnaby, multiple warheads are likely to raise the present U.S. Soviet arms level from 25,000 megatons to 3 million megatons by 1980.

The accuracy of these weapons has also experienced a steady improvement. Andrew Wilson, Defence Correspondent of the *Observer*, has stated that today's missiles have a possible error of one kilometre at 5,000 miles range, but those of 1975 are expected to be accurate to within only 30 yards.

'This means,' writes Wilson, 'that existing forces of land-based missiles, such as the American Minuteman, will cease to be safe from attack in their underground silos. In other words, a situation could arise in which one side might be stampeded into using such rockets for a first strike.'[4]

Some years ago, forced by powerful pressure in their own countries and reinforced by world public opinion exercised through the much-maligned United Nations, the United States and Russia pledged themselves to stop atmospheric nuclear tests which by that time were causing dangerous radiation levels in the earth's atmosphere, posing a threat to the genetic stability of future generations. But it is obvious that it will need a much more concentrated world popular pressure to persuade the leaders of the United States, of Soviet Russia and of China to submit nuclear armaments to effective international control.

To conclude this chapter the following can be said. When the Curies in France discovered the benign effects of radium on cancer, they were not aware of the fateful chain of further discoveries which their pioneering work would eventually produce. This occurred in spite of the fact that Mme Curie herself had become a victim of incurable radiation burns. Neither could Rutherford in England, Neils Bohr in Norway or Albert Einstein in America, all considered as geniuses and mental giants in their time, foresee the disastrous possibilities which their experimenting with the atom would bring upon unsuspecting humanity.

The same applies to such initiators of the industrial revolution as Watts, Faraday, Edison, and a score of others, who were and are still considered benefactors of humanity, while the results of their work are now expressing themselves in environmental pollution and ecological disaster. Only one of them, the Swedish inventor Nobel, who developed dynamite and other explosives, seems to have guessed the fatal consequences of his contribution to science. To atone for it he left his fortune to the Nobel Foundation which was to encourage scientific and literary work for the material and spiritual benefit of humanity. Ironically enough, some of the prizes given in good faith have been for work which in its effect has contributed to the ecological disaster by which we are now faced.

A sober assessment of the present situation reveals that of all the dangers which threaten mankind, nuclear war is the most immediate and formidable one. Beside it the population explosion and environmental pollution appear relatively minimal, at least in their more immediate effects.

In the story of the sorcerer's apprentice, the guiding analogy for this chapter, the little apprentice has been watching the mighty magician. He has seen how the magician ordered his broom to do all kinds of jobs. One day the magician, before leaving his cave, orders his apprentice to fetch water from the well above. When the magician has gone the apprentice immediately tries the magician's formula and orders the broom to take two buckets and fetch water from the well. When the broom emerges with the full buckets of water he empties them into the cave and goes to get more. The sorcerer's apprentice tries to stop the broom but is unable to do so because he does not know the magic formula. So the broom continues to carry water and the water begins to rise in the cave.

The situation in which mankind finds itself at present is in many respects comparable to that of the apprentice.

The sorcerer's apprentice is huddling in the sorcerer's cave, and is already half submerged in water. Meanwhile the broom carrying the buckets lustily hurries up and down the stairs, with ever-new torrents of water which it empties into the cave. Facing the prospect of eventual submersion under the water, the apprentice is straining his brain to find the magic formula which can stop the broom from bringing more water. And here emerges the vital difference between the story of the sorcerer's apprentice and human reality. In the story when almost all is lost, the magician himself appears and calls the magic formula which stops the broom. In the human situation there is no hope for such a magician to appear. Man must find the magic formula himself to stop the follies of his super-brain. Whether he can do so will be the supreme test of his much over-estimated intelligence.

16

The Esoteric Roots of Religion and Philosophy

Such is the importance of speech to mankind that we now propose to consider its nature and development in some detail. This specifically human property has resulted from evolutionary changes in the anatomy of the larynx. It led, in the first place, to improved communications between individuals and, in the second place, through writing, to an accumulation of knowledge culminating in the massive achievements (as well as the drawbacks) of science and technology. Moreover, the mental process involved in articulate speech and writing has led to the exclusive human capacity for forming abstractions. Such language abstractions have been of immense aid to rational thinking, but by some quirk of history they have also led men into the realm of the irrational, with many unfortunate consequences for humanity.

ORIGINS OF SPEECH AND THE MISUSE OF ABSTRACTIONS

We can perhaps illuminate this question of abstractions by taking the simile of the 'smiling cat' from Arthur Koestler's book *The Act of Creation*. Koestler makes the point that 'the grin on the face of a cat' presupposes that there must be a cat and that 'the grin of the cat' when separated from the cat becomes an absurdity and can have no existence of its own.[1] But, of course, there is a sense in which it can enjoy an independent existence in so far as 'the grin' can be presented apart from the 'cat' in its written sign picture or in its spoken tone picture. But 'the grin' in this detached form is pure abstraction, and can be placed in the same category as such abstractions as 'god', 'soul', 'spirit', etc., which also once had physical connections from which they were abstracted and later separated.

How then did the process of abstractions start? Abstractions are part of our language heritage and to explain them we have to examine how language originated.

A rudimentary capacity of speech, the means to communicate by sound and other media, is already present in many animal species. For instance, odours produced from an excited bee's scent-gland, or vibrations produced in a termite colony through the head-rapping of excited individuals on gallery walls, are communicative events. The flashing of fireflies, the chirping of crickets, and the colour changes, movements, and sounds of mating animals are by-products of excited behaviour which are produced to inform other individuals and to alert them. In groups of birds or mammals, the excited cries of one individual will throw others into flight. Barking and biting in a dog change into bared fangs and growling as a means of intimidation. In the everyday life of mammals, we find the 'begging' of hungry dogs and the mewing of cats before closed doors. Lockly reports an example of a pair of sea-horses which were separated in adjoining tanks in an aquarium. Whenever the male made clicking noises, which could be heard in the room, the female would click in response.[2]

Linguistic experts have noted 400 different sounds made by dolphins. There are words in dolphinese to express fear and danger, satisfaction with food, or to call for help. These words seem to be understood by dolphins of different species and areas. It is believed that it may even be possible to compile a basic vocabulary for such concepts as hunger, friendliness, light and darkness.[3]

One of the difficulties of such studies is that dolphins and other cetaceans make sounds with a frequency range of 150,000 to 300,000 cycles, which is far beyond the perception of the human ear. S. Crawford and H. Nissen have shown that chimpanzees can use gestures as language symbols. For example, a chimpanzee will tap on the shoulder of a lagging companion when he needs his cooperation in obtaining food. There is nothing in the structure of their lips or larynx to prevent them from developing articulate speech.[4]

Von Bonin believes that it is difficult to say at what period in the human past articulate language appeared:

> Australopithecus, our early ancestor, certainly walked upright, as the pelvis shows clearly, and one may suppose that by that time the larynx (upper wind-pipe) had gone far enough into the neck to make the enunciation of articulate sounds possible. The jaw, on the other hand, appears still to have been very massive and not well suited to that freedom of movement for the tongue that appears necessary for the production of articulate sounds.[5]

Professor Elliot Smith has remarked that a mould of the brain case of Peking Man reveals a very marked expansion of an area corresponding in position to that part of modern man's brain which is concerned with the appreciation of spoken language.

THE EMERGENCE OF ARTICULATE SPEECH

Human speech abilities, like everything else in organic evolution, have emerged by slow gradations from the less articulate sounds of our animal ancestors. Natural selection will have favoured those family groups of our earlier ancestors which, by means of better language communications, had an improved chance of survival. But variations of speech ability and different languages and dialects are not the direct result of natural selection, but are subsequent developments arising from purely ethnic and historical circumstances.

One can certainly agree with Gordon Childe's remarks that 'words have meaning and suggest things and events only in a society and by tacit agreement among its members'. He continues:

> One member can tell his fellow what he has seen and done, and all can compare their actions and reactions. Thus the experiences of the whole group can be pooled. What parents impart to their offspring is not simply the lessons of their own personal experience, but something much wider: the collective experience of the group. This is the tradition which is handed on from generation to generation, and the method of its transmission with the aid of language seems a peculiarity of the human family.[6]

E. B. Tylor, an anthropologist of the last century, contributed a great deal to our understanding of the origin of words. He writes:

> An example of how word-compounding occurs in unfamiliar tongues can also be found in the Malay term 'arrow', which is *anak-panah*, or 'child-(of-the)-bow'; and the native Australian term for 'unanimous', which is *gurdugrynyul*, or 'heart-one-come'. The Mandingo word for 'sister', *mbading-muso*, shows how such compound words become shortened. It is made up of *mi bado dingo muso*, meaning 'my-mother-child female'. The natives of Vancouver Island gave to a certain long-bearded Englishman the name *Yakpus*: this seems to have been derived from *yakhpeku-kusel-kous*, made up of words signifying 'long-face-hairy-man'. This had been cut down to *Yakpus* in speaking. It struck the Australians when they saw a European book that it opened and shut like a mussel shell and they began accordingly to call books 'mussels' (Muyum).[7]

In the Bornu language of Africa, *tando* to 'weave', has become a general verb to 'make'. In Hebrew *bara* to 'cut', was used to describe the making of the heavens and Earth. Equally our words to make may have meant originally to fit, or join. The English word *sort* comes from Latin *sorts*, a 'lot', through such a set of meanings as allotment, cradle, fate, condition, chance, portion;

kind meant of one kindred or descent; to *be* may have meant to grow; to *suffer* meant to bear burden.[8]

The Greenlander's *uvanga*='I', *ivdlit*='thou', are plainly derived from *uv*='here', *iv*='there'. A different device appears in Malay, where *amba*='slave' is used as a pronoun 'I', and *tuwan*='Lord' as a pronoun 'thou'. How this happened is shown by Hebrew, in such phrases as are translated in the English Bible, 'they servant saieth', 'my lord knoweth'; such terms are on the way to becoming mere personal pronouns meaning 'I' and 'thou', and this actually occurred in the Malay language.[9]

Thus it can be shown that abstract words are formed by abstracting them from real, comprehensible things or situations, so that in the pictorial sense they can be grasped by our senses as if with our fingers.

Gordon Childe writes:

> It must not be supposed that ancient societies were interested in infinite length or empty space. Their abstractions were limited by practical interest. The ancient Sumerian measures of area have in some cases the same names as measures of weight; in particular the smallest unit in both 'tables' is the Se or grain. In other words, Sumerian 'square measure' was originally a seed measure. The Sumerian's interest was the quantity of seed needed to sow his field. He regarded the field not as occupying so much empty space but as needing so much seed-grain. With the areas of bits of uncultivable desert or blue sky he was not concerned.[10]

One must presume that before writing was used prehistoric man had no means of creating abstractions in any way connected with supernatural phenomena such as god, soul, or spirit in the metaphysical sense. The historical situations in which supernaturally linked concepts could have been abstracted from real living phenomena had simply not arisen.

Man originally developed language to describe his physical environment in terms of his sense impressions and he communicated such knowledge to his fellow-men. But this was only possible if all shared the same meaning of words, the same language. For example, if one person points at a cow and says, 'This is a cow' and the other points at it and says 'This is a rabbit', then they obviously do not share the same meaning of language. They cannot communicate in a sensible way; just as there can be no agreement or communication about certain colours between a colour-blind and a normally sighted person.

It can thus be shown that in the process of language development certain abstract terms became denominators which were abstracted from physically existing entities. They were originally used to describe real things by their qualities or attributes, like light, heavy, vaporous, hard, loud, beautiful, the numbers one, two, three, etc. To be comprehensible each attributive term had to be attached to the things it described, and such a description must inevitably have been linked to mutually shared sensory experiences, otherwise they could not have been understood or comprehended by other persons.

Most religions are based on abstractions whose material basis has been lost in the remote past, and the same applies to many philosophical systems. In both cases abstractions are being used in separation from a physically perceptible basis which formerly existed. They have assumed an existence entirely of their own and become divorced from reality. But detached from physical phenomena, religious and philosophical abstractions cannot make sense and must remain purely a play of words or signs. In other words, it is as if one speaks of 'the grin of the cat' as an independent entity, forgetting that there has ever been a cat.

Thus we see that abstractions can only make sense if they are based on real, comprehensible things, or related to physically perceptible situations. In other words, if they are to be genuinely understood they must be related to sensory experience. This principle has been entirely abandoned in the field of religion and abstract philosophy. The resulting aberrations are dealt with under the following headings.

EARLY RELIGION

When thinking about the theory of abstractions it remained for a long time a mystery to me how the tangled events of history can be fitted into the language jungle of religion and abstract philosophy. The idea of a solution occurred when reading J. W. Perry's book *The Origin of Magic and Religion*. In it Perry shows the historical circumstances that can transform everyday occurrences of a rational and apparently scientific order into supernatural abstractions. Perry's thesis, which he developed in association with Professor Elliot Smith, resulted from a study of ideas prevalent in pre-dynastic Egypt.

The Egypt of those times was essentially the Nile valley, which was probably a swampy area stretching along the river bordered on both sides by barren desert. The presence of grain in the stomachs of Egyptians excavated from pre-dynastic graves indicate the presence of early agriculture. This agriculture was essentially dependent on the Nile, in spite of the fact that some archaeologists have assumed that rain was then more frequent than it is now.

The presence of ancient irrigation systems shows that from early times attempts had been made to distribute the flood waters of the Nile over a wider area and thus increase the cultivated land; and this again must have led to a substantial increase of the Nile population. It can therefore be assumed that the maintenance of an irrigation system was of vital importance for the Nile community and that those who controlled and regulated it found themselves in a powerful position.

Perry and Smith contend that the control of irrigation stemmed principally from the knowledge certain people had regarding the exact time of the annual arrival of the Nile flood. They could then issue directives for the next planting. They also believe that such advance knowledge was based on the observations of the moon in connection with the elaboration of a lunar calendar. As the Nile flood appears regularly after an identical number of moon phases (i.e. every lunar year) it was thought that the moon was somehow in control of the life-giving waters of the Nile. Furthermore it was also possible to connect the cyclical changes of the moon with the monthly menstrual periods of women, thereby establishing a causal relationship between the Nile water, the moon, and the life-producing functions of womanhood.

Another incidental relationship resulted from the similarity of the female vulva and the grain of barley which formed one of the principal sources of nourishment. Both vulva and cereal grain resembled the cowrie shell which had been in use since Aurignacian times. From this apparent demonstration of causal relationships, the existence of celestial influences over mundane affairs was demonstrated. And this deduction was strengthened when it was realized that year in, year out, the star Sirius appeared regularly in the East, exactly at the time when the Nile flood began in Middle Egypt.

Professor Elliot Smith has pointed out that the coincidence of the rising of Sirius with the regular innundations of the Nile was believed by the Egyptians to have the relation of cause and effect. It played an essential part, not merely in corroborating the hypothesis of celestial control of human affairs, which the moon's cycles had previously suggested, but also in helping to build up a comprehensive theory of the regulation by the sky of all the really vital affairs on the earth. He maintains that this archaic form of astrology was destined to exert a far-reaching influence upon thought and speculation for the next sixty centuries.

Historical evidence suggests that the individuals who began to play a leading part in the application of such knowledge attained a dominant position within the Nile community in which they acted initially as some kind of astronomer-priest. At a later time they had the status of kings. The early Egyptian kings exercised central control over the water supply of the Nile which was of crucial importance to the life of the Nile people. In this way they practically held power over life and death. Kings came to be considered as the givers of life and the benefactors of the Nile population; eventually they were deified.

Smith and Perry have cited historical evidence dating from the fourth millennium B.C. about a king 'Osiris' on whom the prerogative of an immortal God was bestowed. However, Osiris only became a God after his dead body had been embalmed. This was accomplished by ceremonies which were designed to reinfuse into his body his original sensory perceptions which were thought to have escaped when he died.

The same ceremonial was repeated for future Egyptian kings: each king when crowned was named the son of Osiris, and when he died and was mummified he was identified with Osiris himself. A later development and accompaniment of Egyptian mummification was the making of a portrait statue of the deceased king which, like the mummy of the dead king, was animated by what can be termed a magical process. J. W. Perry says:

> Libations of Nile water were made with the aim of restoring the bodily fluids of the deceased. Then incense was burned to restore the odours of the living. Finally was performed the most important ceremony of all, that of the ritual opening of the mouth, in which a priest touched with a copper chisel the mouth, eyes, ears and nose of the statue, which was then supposed to be alive in the fullest sense of the term.[11]

Perry has suggested that the Osiris cult was the fountainhead from which all future religious ceremonials are more or less derived:

> The priests who conducted the re-animating ceremonies were the pioneers of a long series of ministers who, for more than fifty centuries, in spite of the endless variety of details of their ritual and the character of their temples, have continued to perform ceremonies that have undergone remarkably little essential change. Though the chief functions of the priest as the animator of the god and the restorer of his consciousness have now fallen into the background in most religions, the ritual acts (the incense and libations, the offerings of food and blood, and the rest) still persist in many countries; the priest still appeals by prayer and supplication for those benefits, which the Proto-Egyptian aimed at securing, when he created Osiris as a God to give advice and help.

During the last fifty centuries the meaning of the terms 'God' and 'religion' have become so complexly enriched with the glamour of a mystic symbolism that the Proto-Egyptian's conception of Osiris and the Osirian beliefs must have been vastly different from those implied in the words 'God' and 'religion' at the present time. Osiris was regarded as an actual king who had died and had been reanimated. In other words, he was a man who could bestow upon his former subjects the benefits of his advice and help, but could also display such human weakness as malice, envy and uncharitableness. Much modern discussion completely misses the mark by the failure to recognize that these so-called 'gods' were really men, equally capable of acts of beneficence and of outbursts of hatred, and as one or the other aspect became accentuated the same deity could become a Vedic deva or an Avestan Daeva, a deus or a devil, a god of kindness or a demon of wickedness.[12]

Perry believes that the acts which the earliest 'gods' were supposed to perform were not at first regarded as supernatural. They were merely the boons which the mortal ruler was supposed to be able to confer, by controlling the waters of irrigation and rendering the land fertile. He suggests that the Egyptians derived their ideas in the first instance directly from facts of concrete experience, and that these ideas had in their inception no element of speculation or of symbolism.[13]

The explanation of how the image of a god representing a living man became transformed into a pure abstraction, or spirit, without any relation to physical substance is based on the following considerations.

As the original Egyptian practices expanded and spread out from Egypt they underwent various changes, additions and omissions. In many cases they must have become transformed to such an extent that any image of a god (idol or representation) was considered to be imbued with all the powers of a god, without any of the sensory perceptions having been visibly restored into it. When this happened the pure image of the god (or its representation) became endowed with supernatural power, and unattached to any manifestation of material existence (e.g. a pure spirit).

The further development of such abstractions took place when, for example, the Jews apparently lost the physical image (idol) of their god which they must originally have carried around with them in their nomadic wanderings, and the spirit of their God (Jehovah), all powerful, but now invisible, still remained in existence but as a pure spirit without substance. Eventually, Jehovah became transformed into the Christian, nameless God, the invisible world spirit and creator of the universe.

But how did such animation, originally reserved for a living god-king, become universal so that spirits were supposed to dwell not only in ordinary people, but in trees, stones and animals? Perry has pointed out that originally mummification and immortality remained the preserve of royalty and nobles. When the Osiris belief became later intermingled with the sun cult (fifth Egyptian dynasty and after) the royal family went after death to the sky, while the rest of the community went, if anywhere, to the land of the dead ruled by Osiris.

Perry explains: 'When the king dies, his "ba" or soul, which was his breath, went to the sky in the form of a bird. In the reanimation ceremonies the dead body is given back his breath, his life, he now "existed as a person".'[14] We may add that once the strictness of the ceremonials was relaxed, no limit was set for the abstractions of spirit or soul becoming a common property.

Smith and Perry imply that practices like animism, totemism, magic and related beliefs, which can be found on a world scale amongst savages, were mostly inspired in the distant past by the influence of the higher archaic religions which preceded them and of which the beliefs of primitive peoples are a relic. These superstitions and magical practices represent the degenerate remains of those more complicated religions which were originally practised in Sumer and Egypt.

There is no doubt that the ideas of Smith and Perry contain a large speculative element so that it is possible that some of the events outlined did not take place in exactly the manner they envisage. But this does not invalidate the fundamental argument that supernatural beliefs stem from real life events which happened in man's earlier history, and which were quite rational when they were formed. What really interests us are not only the historical events themselves, on which the origin and growth of religious ideas may be based, but the methods they disclose showing us how natural phenomena, by mere rational association, firstly became transformed into supernatural phenomena; and secondly, when they became detached from real life events, assumed a life of their own free from any physical manifestation.

17

The Aberrations of Religion

In dealing with this subject we will restrict ourselves to that aspect of religion which deals with 'faith', the belief in supernatural phenomena.

As with most other fundamental problems of our existence, so in the case of religious faith, natural selection helps us to clarify the issue. For example, let us consider the widely accepted view, fed by a cursory evaluation of anthropological evidence, that a belief in the supernatural is part and parcel of our human nature. In the section of this book dealing with natural selection and evolution, it has been shown that all living organisms, including man, have been shaped in the struggle for species preservation. Therefore, for faith in the supernatural to have become genetically fixed in the biological make-up of individuals, the process would have required the action of natural selection. According to our knowledge of evolution such a process could only have been accomplished by slow, gradual steps. In such a case natural selection would have preserved such biological traits (produced by gene mutations) as were either beneficial for survival or at least were not harmful. We must therefore examine what survival value a faith in the supernatural possessed.

Let us take the case of an antelope facing a hungry lion. Assuming that it was imbued with faith in supernatural intervention manifesting itself in some kind of reliance on divine protection, the antelope, instead of running away, would remain rooted to the ground and would be duly devoured by the lion. The same fate would, of course, await an African hunter, who, instead of relying on his spear, faced a lion armed with an amulet.

In Haiti, an island greatly devoted to the voodoo faith, about one-third of all newborn babies die of infant tetanus, contracted as a result of animal dung being rubbed ceremoniously on the umbilical knot. This and similar customs, based on religious faith, in Africa, South America and elsewhere, have cost the lives of millions. In famine-stricken India, the cow, a sacred animal, will roam about freely and eat the precious crops. But the starving and dying Hindu, for the sake of faith and religion, will not interfere with it nor use it for food.

One could, of course, cite many other examples where faith must lead to destruction of the faithful. Hence we must conclude that in a state of nature, or among semi-civilized people, religious faith or a belief in the supernatural, must be detrimental to survival. Under conditions of civilization the result may not always be so grave and the capacity of faith may not even be directly harmful. But in no case can it be shown that religious faith is biologically beneficial. Religious faith, or the belief in the supernatural can therefore neither be considered as a normal part of human nature nor can it be proved to be hereditary; just as the smoking of opium is not inherent in human nature. But religious faith, like opium smoking, can become habitual. Karl Marx was not very far from the truth when he said that religion is the opium of the people (but then one must add that in some countries the Marxist faith has become the modern substitute for religion).

It is of interest to note that the preceding argument seems to be contradicted by the conventional interpretation of archaeological evidence derived from prehistoric finds as well as from the activities of living primitive people. For example, the earliest traces of what look like ceremonial burials, indicating a belief in the supernatural, date from Neanderthal times (70,000 to 35,000 years ago). One discovery related by Gordon Childe concerns several skeletons at La Chapelle aux Saints (France), each of which lay in a shallow grave dug in a cave floor. Some of the heads rested on stone pillows and graves were placed near the hearth, as if to warm the occupants. The graves were lined with utensils and food. Childe presumes that this is an example indicating magical practices, although he calls it misapplied science.[1]

W. Howells goes so far as to speak of the possibility of a Neanderthal religion being practised during the fourth glacial period. He adds, however, that the Neanderthalers may have acted like the South African bushmen who inter all a dead man's possessions with him because they want to get rid of them, believing that they are contaminated by his death.[2]

Let us try to pry into the yet uncontaminated mind of a truly primitive man. He has seen many people die and all he knows from actual experience is that it was the end of the life of the person concerned. There was no evidence to show that the dead man (or woman) would have benefited in the slightest degree by providing the grave with food or tools. Further, truly primitive people can never have personally observed that placing a dead body near a fireplace brought it to life again. On the contrary, the stench would have become unbearable. There must therefore be quite a different explanation for the burials of prehistoric people.

One motive that can perhaps be read into the original cause of burials is grief at having lost someone dear, so that some burials may have started for purely sentimental reasons to keep the memory of a loved one alive. For the same reason people may have been buried with their most treasured possessions and their favourite food. There are many animal cemeteries in Europe and the U.S.A. A newspaper clipping from *Time* magazine shows an animal grave with a tombstone bearing the inscription 'Checker 1952-1964 Nixon', explaining that it is the grave of one of President Nixon's dogs. Though Nixon is known to be religious, it can be assumed that he buried his pet not because he believed that it had a soul and that he hoped to meet it again in the hereafter, but out of sentimental attachment – just like many other animal lovers who also bury their pets.

The Cro-Magnons not only buried their dead but adorned them with necklaces of shells and fossils and coloured them with red-ochre. One of the Cro-Magnon tombs, for instance, contains the skeletons of twenty bodies, all flanked with carefully arranged mammoth bones. Aurignacian cave paintings have been found in various parts of the world, from the Pyrenees to Russia, and from Palestine to South Africa. Many of the representations discovered show exquisitely executed hunting scenes, some in colour, placed deep in the recesses of caves. In addition these early people left decorative carvings in bone and stone. Amongst them are numerous mother figurines, often with their sexual parts greatly exaggerated.

It has been pointed out that the Aurignacians depicted hunting scenes for magical purposes in the belief that this could increase their food supply. There is, however, no evidence to support such a view. As a matter of fact, when closely examined, such deductions are completely in conflict with ordinary common sense. For example, no Aurignacian can ever have observed that cave pictures increased his food supply in the slightest degree. On the contrary, if the people who made these paintings had helped in the hunt more food would probably have been provided. Furthermore, if we take the question of painting in general and ask for the motive which induces ancient as well as modern painters to produce their work, we will find hardly any example where the pictures were or are produced for magical purposes. Why then should the Aurignacians be attributed with such odd sentiments? Therefore, until a better explanation comes along, we must consider prehistoric cave paintings as mere representations used to commemorate significant events in the life of a hunting people, and strip them of any magical or religious context.

Mother figures are another case of apparently magical pursuits.

As mentioned before, they have been found in many parts of the world from Aurignacian times onwards. Recently, this subject was studied in detail by Peter Ucko. His conclusions are that such figurines do not suggest any worship of fertility goddesses, or the use of magic.[3]

'Magic,' writes Professor Elliot Smith, 'is nothing more than the scientific methods of the past which have been shown to be unsound, but survive as popular superstitions.'[4]

One rationalization which has been put forward to explain the belief of primitive people in the supernatural is the contention that they must have felt rather helpless when faced with the inclemencies, as well as the wonders, of nature. It is assumed that they were so impressed by the awful force of tropical thunderstorms, by the eruption of volcanoes, by such events as rain and sunshine, and by the miraculous fertility of the soil, that they believed, as a matter of course, that all these phenomena were the work of powerful spirits and ghosts that abound in nature.

But a closer study of primitive people, as well as of ourselves, indicates something else. Anthropologists like Lévi-Strauss, who have lived for long periods amongst primitive people, have repeatedly expressed their amazement about the wonderful common sense these people show in their everyday pursuits. But such rational behaviour must be kept apart from the many traditional, irrational customs and beliefs which primitive people often hold, and which they have inherited from their ancestors. In studying their behaviour it must never be forgotten that primitive people are on the same biological intellectual level as ourselves. They must be weighed with the same measure of understanding and the possession of scepticism minus their traditions and prejudices which they (and we) have inherited from predecessors. Naturally, it is difficult to trace the origin of most of the irrational superstitions and traditions which we, as well as many primitive people, still retain. But we can be sure that the original idea must have been quite rational and only became irrational when the material connection which must have originally existed was lost. Some insight into the mind of primitive people can be gained from the observations of J. E. Griffith and J. W. Perry on the Ponde tribes of South Africa. Both these ethnologists were for several months in contact with an old man named Mtandama. On one occasion Perry gave him an accurate version of the creation story of the Huron Indians of America (according to this story some of their early ancestors lived in the sky. In the distant past they dived from the sky into the depth of the ocean and brought up some earth. Then all the people went down to settle on it.) Perry also told Mtandama that the Hurons believed in a sky-world. This went on for many days, but failed to make any impression on Mtandama's matter-of-fact attitude towards other people's abstract ideas. 'In his mind, the sun was just the sun and the moon the moon. They remained the same as he had known them from childhood and he had never taken any special notion of them.' Perry asked him if he would like to go up and see the Victoria Falls. He replied, 'Why should I? It's only a lot of water, and there is plenty down here to look at.'[5]

Similar instances were reported by Professor Malinovski about the Tobriand Islanders. He says the natives are aware of the sky and the movement of the sun and its influence on the fertility of the soil, but they have never used this knowledge to form the idea of a solar year or a calendar.

'When pressed,' writes Malinovski, 'the natives will say that the sun walks or moves across the sky, that it dips down at sunset, moves under the rim of the earth and rises again on the other side in the morning. But even this explanation would only be advanced by the most intelligent native without the least enthusiasm or interest. Any other question would be answered by "anayseki" (I don't know).'[6] Malinovski also reported that although the moon plays a much bigger part in the life of the natives than either sun or stars, there are no traces of the pseudo-scientific curiosity frequently attributed to natives.

The facile assumption, therefore, of some anthropologists and psychologists that our forefathers formed senseless and childish notions about nature merely out of the blue must therefore be rejected. Let us take, for example, blood. Nobody denies that it may be within the intelligence of primitive people (as it is in our own) to observe that the excessive loss of blood can cause death to man or animal. But that red-ochre (as used by prehistoric man) can spontaneously be considered as a substitute for blood is certainly entirely illogical. Because no one can ever have observed that by substituting red-ochre for blood a dead body could have been brought to life again. To account for the use of red-ochre in the preparation of corpses or other apparent magical pursuits, from Mousterian times onward, an entirely different explanation is called for.

The contention therefore that the belief in supernatural powers has been spontaneously created by primitive people must be considered pure conjecture and is not supported by known archaeological or anthropological evidence. This brings us back to the argument that the belief in the supernatural must have come about quite accidentally when, through attempts of early science, certain associations between mundane and cosmic events were formed by pre-civilized people. It is possible that this happened when the attributes by which natural phenomena can be described became detached from their formerly physical basis. Although, having thus become separate entities, the abstractions thereby formed survived and eventually assumed a life of their own. This was made possible by what can only be described as 'a perversion' of everyday language. Relatively speaking 'the grin on the face of the cat' assumed a life of its own, separate from any manifestation of a real cat.

18

The Pitfalls of Abstract Philosophy

Whatever may eventually come to light about the true historical origin of religious beliefs, it is certain that the ancient populations living around the Eastern Mediterranean, including the Greeks, derived their early religious and scientific ideas from Egypt and Mesopotamia. It was only later (from about 600 B.C. onwards) in their re-interpretation of these ideas that the Greek philosophers played a part which has remained of lasting impact even in modern civilization.

It is suggested here that what the early Greek philosophers actually tried to accomplish was to endow the religious ideas with a rational content. As it turned out, the Greek philosophers failed to solve the ancient mysteries. What they achieved was merely to replace the mystique of these archaic religions by their own, equally mysterious system of abstract philosophy. In this process, religion, abstract philosophy, and true science became intermingled and this concoction has influenced the mental life of mankind ever since.

Arthur Koestler refers to this process in his book *The Act of Creation*, showing that the principal heroes (or villains) of the drama were Pythagoras, Plato and Aristotle. Koestler writes:

> The record starts in the sixth century B.C. when we find suddenly, as if sprung from nowhere, a galaxy of Philosophers of Nature in Milos and Elea and Samos, discussing the origins and evolution of the universe, its forms and substance, its structure and laws, in terms which have become forever incorporated into our vocabulary and our matrices of thought. They were searching for some simple, ultimate principles and primeval substances underlying all diversity; four elements, four humours, atoms of a single kind, moving according to fixed laws. The Pythagoreans attempted the first grand synthesis; they tried to weave the separate threads of religion, astronomy, and music into a single carpet with an austere geometrical design. That carpet is still in the making, but its basic pattern was laid down in the three centuries of the heroic age of Greek science between Thales and Aristotle.[1]

Connections between Greece and Egypt are indicated by Herodotus. He reports to us that the Greeks derived their ruling families and gods from Egypt. Danaus is supposed to have been the son of an Egyptian king and the nephew of the king of Phoenicia. Heracles, hero of Greek mythology, also came from Egypt where he was worshipped as one of the Egyptian deities. On the other hand, Greek mathematical and astronomical ideas were mostly derived from Babylonia.[2] Pythagoras' theorem is not an invention of Pythagoras himself but was rediscovered by him through his contacts with Babylonia where it had been known as early as 2,000 B.C. 'In fact,' says Gordon Childe:

> We do not know how any of the geometrical rules were obtained. They certainly had not been deduced *a priori* to be in Euclid's geometry. For such a 'pure geometry' there is no evidence at all. Figures often accompany the geometrical problems in the mathematical papyri and tablets, but such figures are no more drawn to scale than are the field-plans in contracts. On the other hand, the patterns decorating craft-products, piles of bricks, and composite wooden boxes often give very striking colour demonstrations of geometrical propositions ... Fourteen hundred years later the physical origin of such geometrical abstractions as well as those of numbers and mathematical formulae had become obscure and forgotten. The mystery relationship between squares and circles, between squares and triangles, etc. seems to have been taken over by Pythagoras, as well as Aristotle, Plato, and others as philosophical truths.[3]

On the other hand, Herodotus (in the *Histories*) reports: 'It is said that Pythagoras learned numbers and measures from the Egyptians; astonished at the wisdom of the priests, which was deserving of belief and full of fancies and difficult to buy, he imitated it and himself also taught his disciples to be silent, and obliged the students to remain quietly in rooms underneath the earth.'

Another connection between the archaic religions of Egypt and Sumer and Pythagorean ideas arises from the relationship between the Nile flood and the star Sirius. Because the star Sirius is seen on the horizon at dawn when the Nile flood arrives, it was inferred that Sirius caused the Nile flood. This event not only led to ideas of astrology but induced Pythagoras to proclaim that the human soul was originally a star.

According to Aristotle (*Metaphysics*) Plato's doctrine of form was, in its general characteristics, not different from Pythagoreanism, the forms being actually called numbers.[4] Making the best use of the science available at the time, Plato undertook to show that the principal movements in the cosmos, including that of the sun and planets, are in fact perfectly regular. Given the principles that he had just stated, it follows that the world is controlled by divine beings, perfect souls, to whom we are akin. According to

Aristotle growth and change in the physical world are to be explained in analogous fashion. This conceptual analysis is then applied by Aristotle in the following way. God alone is fully actual, a form without matter, who does not have to be raised from the potential to the actual state. His activity is the uncaused sense of all change; but he is its 'final cause' or goal and he does not, as in the Platonic system, design the world.

There is much evidence to show how early Greek philosophy influenced the mental life of later ages. Plato's Academy is the link with the arithmeticians of Alexandria. Furthermore, Plato's idea that there is another more intelligible world, more real than that with which we are sensibly familiar, has been absorbed into the Christian tradition.

Pythagorean practices and doctrines, particularly that on immortality, are attested in Rome from the first century B.C. As a philosophical theory Pythagoreanism merges with Platonism. But it was also a kind of cult; and the frescoes in the underground basilica of the Porta Maggiore have been claimed as Pythagorean. In imperial times there were Pythagorean philosophers and mathematicians in Alexandria, in Syria and elsewhere. Plato's immediate successors had already made a partial return to more purely Pythagorean theory; and neo-Platonists, such as Lamblichus in the fourth century A.D., drew from them as well as from more recent forged Pythagorean writings (e.g. the Orphic Hymns).

Arab philosophy between about A.D. 990 and 1200 is also linked to Aristotleanism and neo-Platonism. It can, for example, be shown that the ideas of the Arab philosopher Al-Kindi (about A.D. 990), Avicenna of Persia (A.D. 979-1037), and Averroes, a Spanish Arab (A.D. 1126-1198), rest on Aristotlean analysis. For Christian Fathers, such as Ambrose, Pythagoras was an authoritative figure because he was thought to have been a Jewish intermediary between Moses and Plato. By the sixteenth century he was liable to be cited, according to the interest of the writer, as a poet, as a magician, as the father of the Cabbala, as a mathematician or as a champion of the contemplative life.

It would be idle to expand here further on the ideas of later philosophers like Descartes, Spinoza, Leibniz, Kant and others (excepting of course their scientific and mathematical achievements.) In spite of their apparent differences they all deal in pure philosophical abstractions which they endeavour to give an existence of their own. But just as it is not possible for the 'grin of the cat' to have a life of its own, while the 'cat' is missing, philosophers cannot make sense of something which cannot be sensed. At the same time it is not denied that the purely scientific achievements of some of these philosopher-scientists are of the highest intellectual order and of practical application, but their philosophical abstractions were not.

But we are concerned here with abstract philosophy and religion and not with science. No one can, for example, be unaware of the tremendous influence religious ideas have had on human history. And although today their influence has diminished, it is still considerable. Of the influence of abstract philosophy, on the other hand, we can say that up to the middle of the last century its main impact consisted in modifying religious thinking by making it more refined and sophisticated.

Only one philosophical system can claim to have succeeded in exercising a decisive influence on human affairs. During the last century the abstract philosophy of dialectical materialism has become the basis of scientific Marxism, which has succeeded in bringing under its mental and physical domination a large proportion of the human race. This does not mean that the broad masses of the communist world know what it is all about, but for the dominant ruling minorities it is the philosophy from which they derive their power. In view of the role Marxist philosophy plays in the world today, it is important to take a look at its basic principles. For this reason we have to go back to the science of logic and dialectics.

LOGIC AND DIALECTICS

Logic (meaning the science of logic) as defined in *Encyclopaedia Britannica* 'is the systematic study of the structure of propositions and of the general conditions of valid influence by a method which abstracts from the content or matter of the propositions and deals only with their logical forms.'[5]

It can be noted that this definition deals entirely with the 'grin of the cat', with the 'cat' nowhere in sight. The pure abstraction, 'logic', is explained by means of an attribute of the same abstraction, 'logical pursuit'. The simple fact remains that there is no logic in *pure* language, apart from the mere play of words; neither can pure logic be found in mathematics, except in the game of bare numbers. To be comprehended and practically utilized both these departments of knowledge must be related to real physically perceptible phenomena. The conclusion is that there is no logic (i.e. if translated as common sense) in logical science or in abstract philosophy.

The English philosopher, Bertrand Russell, has generally been recognized as the modern champion of the science of logic and it has been said that the greatest achievement of his pursuit into logic was his *Principia Mathematica* (written with A. N. Whitehead). Yet in his memoir *Reflections on my Eightieth Birthday*, Russell confesses that the mathematical structure he had worked so hard to erect was nothing but an illusion.

He writes, 'I set out with a more or less religious belief in a Platonic eternal world, in which mathematics shone with a beauty like that of the last Cantos of the Paradiso. I came to the conclusion that the eternal world is trivial and that mathematics is only the art of saying the same thing in different words.' Russell's late reappraisal is amply confirmed by an analysis of 'dialectics'.

In the definition of *Encyclopaedia Britannica* (1961 edn.) 'dialectics' is a part of logical science. Aristotle is not only supposed to have been the founder of the science of logic but was also the first to use dialectics, both being subjects closely associated with each other.

The Aristotelian view about dialects can be traced to his *Metaphysics*. Elements of numbers themselves were 'the odd' and 'the even', or 'limit' and the 'unlimited', which in Aristotle's view represent a primitive insight into the notions of form and matter. These two pairs headed a list of ten pairs of fundamental 'opposites', the remaining eight being 'one' and 'many', 'right' and 'left', 'male' and 'female', 'rest' and 'motion', 'straight' and 'curved', 'light' and 'darkness', 'good' and 'evil', 'square' and 'oblong'. It was a philosophy of metaphysical and moral dualism; but it came to see the universe as a harmony of opposites, in which 'the one' generating the number series, or 'limit', successively imposed itself on 'the unlimited.'[7]

There is a short analysis of Aristotelian opposites in the Appendix (p. 172-4). The conclusion (as the reader will find) is that the play of opposites turns out to be pure fantasy without scientific or any other significance.

'In the system of Hegel,' writes the *Encylopaedia Britannica*

> Aristotlean dialectic becomes the tendency of a finite object or notion to 'suppress itself', to develop contradictions and to pass over into its own negation. Such dialectic occurs in the realms of thought and science, but it also occurs in the spheres of nature and history. Wherever there is movement, wherever there is life there dialect is at work.

The Hegelian word and notion were to some extent taken over by Karl Marx. Hegel believed that thinking always proceeded according to this pattern: it begins by laying down a positive thesis which is at once negatived by its antithesis; then further thought produced the synthesis. But this in turn generates an antithesis and the same process continues once more, but not indefinitely, for it is circular. Ultimately thinking reaches a synthesis which is identical with its starting point, except that all that was implicit there has now been made explicit.[8]

The interplay of Hegelian and Marxist opposites is also very similar to the dialectics expressed by Heraclitus of Ephesus (about 500 B.C.). Heraclitus says, 'To identify any one thing is also to name its opposite; for day and night are one ... The process of coming into being and passing away is a constant intertransformation of opposites, one into the other.'

Let us now consider the effects of dialectics on contemporary history. An interesting comment on Marxist and Hegelian dialectics can be found in the writings of Stalin. He writes: 'Marxism is not only the theory of socialism, it is an integral world outlook, a philosophical system, from which Marx's proletarian socialism logically follows.'

'This philosophical system,' says Stalin, 'is called dialectical materialism. Hence, to expound Marxism means to expound also dialectical materialism. Why is this system called dialectical materialism? Because its method is dialectical, and its theory is materialistic.'

He continues, 'Hegel's philosophical system, which rests on the immutable idea, is from beginning to end metaphysical. But it is also clear that Hegel's dialectical method, which repudiates all immutable ideas, is from beginning to end scientific and revolutionary. That is why Karl Marx, who subjected Hegel's metaphysical system to devastating criticism, at the same time praised his dialectical method, which, as Marx said, "lets nothing impose upon it" and is in its essence critical and revolutionary.'

What Marx did, therefore, was to transform the metaphysical abstractions of Hegel into the equally incomprehensible philosophical abstractions of a so-called materialistic pseudo-science, mixed with historical and economic notions. Marx's application of dialectics to the play of opposites remains substantially the same as that outlined by Aristotle and continued by Hegel. Whether we call the abstract philosophical theory of opposites 'idealistic dialectics' or 'materialistic dialectics', it still makes no sense from a common sense point of view, irrespective of whether it is advanced by Aristotle, Hegel, Marx, Stalin, or their later followers.

Yet this book deals also with practical problems. It is therefore of interest to see what practical effect the philosophy of 'historical and dialectical materialism' has had on the economy of the Soviet Union. The following example is an apt illustration. At the 20th Congress of the Soviet Communist Party in 1956

Russian Premier Nikita Kruschev related the following story.

'Stalin [1953] proposed that the taxes paid by the Kolkhozes and by the Kolkhoz workers should be raised by 40 billion roubles (about 40 billion dollars). According to him the peasants were well off and the Kolkhoz worker would need to sell only one more chicken to pay his tax in full.

'Imagine,' said Kruschev, 'what this meant. Certainly 40 billion roubles is a sum which the Kolkhoz workers did not realize for all the products which they sold to the government. In 1952, for instance, the Kolkhozes and the Kolkhoz workers received 26,280 million roubles for all their products delivered and sold to the government.

'Did Stalin's position then rest on data of any sort whatever? Of course not. In such cases facts and figures did not interest him. If Stalin said anything, it meant it was so – after all, he was a "genius", and a genius does not need to count, he only needs to look and can immediately tell how it should be. When he expresses his opinion, everyone has to repeat it and to admire his wisdom.

'But how much wisdom was contained in the proposal to raise the agricultural tax by 40 billion roubles? None, absolutely none, because the proposal was not based on an actual assessment of the situation but on the fantastic ideas of a person divorced from reality.'

It may be asked what all this has to do with dialectical materialism? The reply is that the above very drastic economic measure was proposed by a man who was utterly convinced of the decisive influence of dialectical materialism on economic affairs and who determined the day-to-day economic policy of 200 million people. At that time, Stalin was genuinely considered as a genius not only by the 200 million Russian citizens but also by the entire world communist movement.

No further comments about the practical side of the philosophy of dialectical materialism need be added, except that a monstrous mismanagement of human affairs is being committed on hundreds of millions of people in the name of some abstract philosophical principle which is entirely irrational. This does not imply an apologia of capitalist economics the results of which are sometimes equally devastating.

To conclude this chapter, we can say this. Natural selection shows us that religion (the belief in supernatural phenomena) cannot be the consequence of any particular, inborn (hereditary) trait, but has resulted from certain historical circumstances, and that abstract philosophy merely resulted from the attempt to rationalize ideas which were originally religious. It seems that supernatural and abstract philosophical concepts arose from a misapprehension of strictly physical phenomena due to language perversion.

Thus, in the first stage of this development, mundane events became associated with heavenly phenomena and were thereby endowed with superhuman powers. In this initial process 'the grin of the cat' became, symbolically speaking, a celestial grin. During the course of history, as some living men became transformed into gods, supernatural attributes became separated from their originally material association while their original association was forgotten. So, such abstracted expressions as 'god', 'soul', 'spirit' and others, assumed a life of their own. This was the stage when 'the celestial grin of the cat' became detached from 'the cat' itself, a pure abstraction.

Then we reach the period when the Greek philosophers attempted to rationalize about the purely abstract ideas of the archaic religions, concerning 'god', 'spirit', and 'soul', and so created a whole new system of abstract philosophy. It is at this stage that the celestial grin of the cat became more earthly, though the cat was still missing. And, finally, we have the attempt of modern philosophy to wrap the abstract philosophical ideas of the Greeks into truly earthly garments and adorn them in the process with some new fancy philosophical frills. The results include dialectical materialism, the abstractions of Freud (the 'unconscious' and the 'id'), linguistic philosophy, modern logical science, and philosophical considerations dealing with mathematics. The grin of the cat has now become earthly but remains still detached from the cat.

This, then, is an assessment of religion and of abstract philosophy from a biological point of view. It has been undertaken with the intention of removing the mental and ideological fog which obscures the true recognition of our biological predicament.

There was a time when the continent of Europe (during the Dark Ages) was politically dominated by the Catholic Church. Under it the free expression of ideas and their interflow was considered a danger to the faith, and non-believers were persecuted. Similar tendencies of intolerance prevailed in the Islamic world.

During the so-called Dark Ages, the scientific and intellectual achievements of the Greeks became obliterated and were only rediscovered and enlarged upon when religious domination receded. This took place during the last three centuries, when both religious systems (i.e. Catholicism and Islam) lost much of their direct political influence. In most countries church and state were eventually separated, religious tolerance was established, and scientific investigation was again permitted, but in those countries where religion has retained a dominating influence, the populations have more or less remained socially and economically backward (witness such countries as Spain, Portugal, Saudi Arabia, Ethiopia and others).

It is nevertheless interesting to note that up to recent times religious dominance exercised such a deep influence that in some states of the U.S.A. (Tennessee for example) it was forbidden to teach Darwin's theory.

The conclusion is that whenever there was religious domination, intellectual darkness prevailed. And while it is admitted that the moral teachings of religion (with the exception of sexual morals) have generally exercised a beneficial influence on human affairs, its other aspects have been detrimental to human biological aspirations.

Now it must be understood that while attacking the supernatural premise on which religion is built, the thesis of this book is not really anti-religious in the sense that it aims at advocating the forceful prevention of religious worship or that of theological discussion. All it attempts to do is to point at the dangers which arise from a domination by religious ideas of the body politic.

The same is the case with abstract philosophy. Orthodox communism has in many ways assumed the mantle of Catholicism. And where in the former case doubters were treated as heretics, doubters of the communist dogma are equally persecuted. How this can affect science has been shown in the Lysenko affair when Stalin tried to manipulate biological science by resurrecting the Lamarckian tenet of the heredity of acquired characters, suppressing the teaching of genetics. Several prominent scientists were put on trial and ended their life in Siberia. There is indeed not much difference between the trial of Giordano Bruno, in 1500, and Stalinist witch-hunts in the twentieth century.

19

The Failure of the Great Religious and Libertarian Movements

Religions

Apart from their theological side, the religions of Christianity, Mohammedanism and Buddhism, each of which counts hundreds of millions of followers, all base their teachings on the unifying force of commonly held moral and ethical principles. They have swept across frontiers, they have broken the barriers of tribe and race, and they have engulfed whole continents. Despite this, they have not succeeded in breaking the archaic order of sovereign states which has ruled the world for thousands of years.

Let us take Christianity. Christ's teachings are based on the loftiest moral principles in which love plays the main part: love thy neighbour, be charitable and succour the needy and helpless, forgive thine enemy, practise the brotherhood of man. But Christ's moral principles are only one part of the Christian religion and Christ is only the junior partner in the heavenly hierarchy. The senior partner is his father, God himself, the almighty creator of the universe. It is said that without God's will no leaf can fall from a tree; no speck of sand can move in the desert. Consequently, as God is all-knowing and all-powerful, he must bear the ultimate responsibility for all that happens on this earth.

In the German language the word God (the Christian God) is preceded by the attribute 'loving' and one speaks always of 'the loving God' ('Der liebe Gott'). This 'loving God', if he is really all-knowing and all-powerful, must be held responsible for all the crimes and misdeeds committed on this earth. In such a context God's position resembles that of a father who gives his little children sweets poisoned with arsenic. The children die under the most painful convulsions. When the father is questioned, he maintains that he cannot be held responsible for the death of his children: all he did was to give them the sweets and it was of their own free will that they ate them.

In the same relationship stand the deeds of God to the free will of men. Accordingly, we must assume that lately God has given his children some toys called atom and hydrogen bombs and awaits the outcome with indifference. In such a context the Christian God (and the Jewish and Mohammedan God as well) must be assumed to be a callous and revengeful creature, utterly insensitive to all human suffering.

What must be obvious by any sane human standards is that it is inconceivable that there can exist such a monstrous personality, whether he is called God, World Spirit or anything else, except as a figment of pure imagination or theology. How the abstract conception of god arose has been explained in an earlier chapter. Yet the image of the revengeful and just god continues to persist in the fantasy of the Jews, the Christians and the Moslems.

Christianity is torn by a terrible and irrational dualism. On the one hand, there is Christ imbued with the highest moral principles. On the other hand, there is God his father who has the power to end all injustice and suffering on this earth. Yet, instead, God crucifies his son. It is clear that such a mixture of morality and immorality cannot assist in the salvation of mankind, whatever arguments Christian apologists may try to produce.

It is not surprising therefore that under the combined teachings of Christ and his God-father the Inquisition tortured and executed countless men, and that thousands of guiltless women were burned alive as witches. In the light of this duality, Christian priests have blessed and are continuing to bless, the arms which are destined to maim and kill not only men of other creeds, but also their own Christian brethren.

There is no doubt that the moral and ethical principles which these religions proclaim are a valuable heritage which should be preserved. But belief in an all-powerful, irresponsible God must be abandoned. Further, the thesis of 'My kingdom is not of this world' must also be rejected. Human morals are based on the flesh, and are related to our sensory emotional and instinctive life. They have nothing to do with any abstract or so-called 'spiritual' principle.

Buddhism, the teachings of Tolstoy, and those of Gandhi, have also failed to change the course of human affairs, in so far as this concerns the eradication of violence and war and other forms of abhorrent behaviour. Although their teachings are based on ethical and moral principles akin to those of Christ, they, like Christ, rejected our sensory and instinctive life. By denying the sublimation of our most profound biological needs, they, like the Christian religion, have created deep fissures in the psychological and emotional balance of individuals and in society. By establishing and encouraging the sexual lie (the attempt to hide and falsify everything that has to do with sex) they have created a basically hypocritical and frustrating society which finally has to find an outlet in paranoia, violent dissent, and blind rebellion.

Thus we find that the followers of Buddha, in spite of their high moral principles, have been waging tribal and national wars in Asia for centuries, just as their Christian counterparts have done in Europe and elsewhere. We also find that the disciples of the non-violent Gandhi became, with few exceptions, involved in the communal strife and massacres in which more than 3 million Indians were killed, even while Gandhi was still alive.

In *The Kreutzer Sonata*, guilt-ridden Leo Tolstoy, driven all his days by strong sexual urges (which he tried to suppress) pronounced his own final solution to man's conquest of the flesh: the proposal of total sexual continence. By suppressing the sexual urge entirely, men would cease to leave any offspring and the human race would quietly expire. With the disappearance of man, sin and temptation would also die out.

Libertarian movements

In the great mental and political upheavals which preceded and followed the French Revolution the notions of liberty, equality and fraternity were born; natural selection and organic evolution were discovered. Man had finally succeeded in breaching the formerly unassailable divine order based on birth, privilege and wealth. Although the French Revolution ended in the dictatorship of Napoleon, things were never to be the same again.

The most idealistic of the libertarian movements emerging from 'the age of reason' was that of anarchism. Anarchists had discovered that man was fundamentally a moral entity and that simple communal life should be the goal. In such a visionary society there was to be no private property and everyone would contribute according to his abilities and partake according to his needs. This free association of individuals would preclude the necessity of law and order. There was therefore no need for the state to continue. Once coercive laws were removed, man's true nature would assert itself and the whole world would live in peace and harmony. Some of the anarchists considered the coercive machinery of the state as the main culprit and, as they assumed that those in power would never give in voluntarily, advocated its overthrow by violent means.

What the anarchists had forgotten was that except in the case of primitive isolated family groups such as a pristine state is an illusion. Within a large community of individuals, living under advanced economic conditions, with a great diversity of interests, certain basic principles of law and order have to be imposed. Otherwise, the unrestrained actions of irresponsible and selfish individuals, who, as we only too well know, exist amongst us, lead to the worst kind of tyranny and subjugation over the majority of meeker and more peaceful members of society.

The great libertarian movements of socialism rest basically on the same premise as anarchism. Both democratic socialism and Marxist communism draw their inspiration from the belief that man is inherently good and perfectible with the proviso that socialism's success depends principally on economic factors. But scientific socialism is also based on the idea of progressive organic evolution with man eventually occupying the top rung of the evolutionary ladder, propelled there by the force of Lamarckian inheritance. In democratic socialism, as well as in Marxist communism, the result is presumed to be inevitable. It is the attainment of the anarchist heaven of a free world community, where each one lives according to his needs and contributes according to his abilities, where man will reach the height of his evolutionary development, mentally, physically and morally, and where there is no need for private property, no need for law, and where people will live peacefully and harmoniously in voluntary association. It is that stage of society in which the state, in Marxist phraseology, will have withered away. Thus the aim is the same as in anarchism, although the methods of achieving it are different. Social democratic movements attempt to reach their objective by gradualist democratic processes; Marxist communism tries to do so by way of violent revolution.

The main reasons why democratic socialism and Marxist communism have failed are these. Firstly, their teachings are based on a wrong interpretation of organic evolution. There is no automatic progress towards organic perfection and the Lamarckian inheritance on which social evolution is supposed to be based has proved obsolete. Secondly, socialist concepts are based on a wrong assessment of economic factors, i.e. on the assertion that economic influences determine all social, moral, and biological relations. But if the human situation is examined from the biological point of view, it can be shown that it is biological factors which determine our economic needs, and moral and ethical factors which determine our social relations. There is no doubt that economic factors exercise a profound influence on all aspects of human life, but the ultimate motive is and remains biological.

As for Marxism, as distinct from democratic socialism, we find a third reason for its failure. It is based on the sterile philosophical exercise of dialectical materialism, a senseless concoction of abstractions, leading to an arbitrary deterministic interpretation of the historical process. But history is based on millions of different incidents affecting millions of different individual lives, and thus creates innumerable unpredictable situations. All we can therefore do is to improvise all the time and remind ourselves that any rigid social or economic system must lead to failure.

In the national struggles for the attainment and retention of political power, democratic socialist and communist movements have invariably acted on the principle of expediency, with greater or lesser disregard for the originally professed moral and ethical principles. We can say that the aims of socialist movements were magnificent, but the basic ingredients on which they depend are faulty, and the results are therefore predictably disappointing. In those countries, for example, where communism has been established (Russia, China, the communist people's democracies) we invariably find a grim picture. The example of Russia reveals a tyranny maintained by a vast apparatus of secret police. On the lower level, the entire population is smarting under a monstrous bureaucracy. It is based on a negation of fundamental human rights, imposition of arbitrary justice, and the repudiation or absence of any moral principles in government.

In China, economy and government are guided by the same esoteric means. Here everything is based on the little red book of Chairman Mao which consists of a number of quotations that, like the sayings of the Bible, of the Koran, or the pronouncements of Lenin, can be utilized to explain and justify almost anything under the sun. The result is again a vast bureaucracy, with the complete suppression of individual liberties.

Yet despite the heritage of Stalin and the tantrums of Mao Tse-tung, the Russian and Chinese people continue to survive, although the political and economic machinery of both countries is creaking under the strains and stresses of doctrinaire rigidity. Reportedly the vast majority are much better off than they were under the oppressive reign of the Tsars or the Manchu rulers. The situation of the mass of the people may be compared to that of inmates of a prison. They have been relieved of their chains and their basic physical needs (i.e. education, health, housing) are being taken care of, but spiritually and politically they are forced to conform to an oppressive ideology.

If socialism and communism are stripped of their doctrinaire trappings, some of the original assumptions still retain their value and need to be preserved. They are, in the moral field, the belief in the basically decent nature of man. And, in the economic field, the necessity for the abolition of economic exploitation and the establishment of political conditions that accord with fundamental human rights. In both these respects socialist aspirations can be seen to coincide with the biological needs of mankind.

Conclusion

Natural selection raises many problems. Foremost among them is its influence on evolutionary theory. As a general rule, an intensive natural selection keeps animal species constant and at a healthy level of biological efficiency. Thus we find a condition which can be described as evolutionary stagnation and which is one of the most common occurrences in animal life.

In man-like species there has been an evolutionary trend towards a larger brain and greater intellectual powers. But this has been paralleled by a weakening of physical structure. In the human species with the invention of fire, clothing and other cultural aids (dating back perhaps half a million years) life became easier and the intensity of natural selection was reduced. This not only made any further evolution of our intellectual powers unlikely, but has led to biological recession (degeneration). So far, this has not posed a threat to our species survival because science and technology have compensated for consequent biological deficiencies.

In studying the implications of human degeneration we recognize that the process is virtually irreversible, though there may be a possibility of arresting and stabilizing it. The fact is that man-made selection cannot equal the effectiveness and impartiality of the natural process. On the other hand, the reimposition of unrestrained natural selection (if that were possible) would require the deliberate abandonment of culture and civilization as we know it. It would mean the return of mankind to a simple food-gathering and hunting life, which, in turn, would call for a reduction of the world's population from its present 3,600 millions to perhaps 50 millions.

A further consequence of human evolutionary stagnation is the dethronement of established anthropological and sociological theories of culture which are linked to biological evolution based on Lamarck. This invalidates the theory that the world's principal cultures and civilizations arose in isolation in some sort of evolutionary parallel development. It also leads to the abandonment of such Lamarck-based ideologies as scientific Marxism and democratic socialism, both of which founder on the rock of natural selection. This puts Marxist countries such as Soviet Russia and communist China and their satellites in a considerable ideological dilemma. Natural selection also debunks the socio-psychological interpretations in ethnology and sociology of Freud and the assorted schools of sociology and structural anthropology to which they are related.

Thus, in the place of the isolationist thesis of cultural anthropology and the Freud-based thesis of psycho-sociology there emerges the aspect of world-wide cultural unity and interdependence. In this new relationship, institutions like divine kingship and the sovereign state, reveal themselves as arbitrary structures of a common historical origin, which can neither claim permanency nor any particular usefulness.

Natural selection also emphasizes the importance of our instinctual life. We find that notions of inborn human aggressiveness turn out to be greatly exaggerated, while, contrary to what is generally supposed, our ethical and moral behaviour is found to be based on strongly entrenched instincts that helped in the survival of the primitive family group.

It has further been shown that our sensory equipment is subsidiary to the instincts and that the coordination of sensory perception with the aid of memory results in the mental process. In the final analysis our mental (intellectual) powers are revealed as the servants of the instincts. Unfortunately, with the development of culture and civilization, intelligence has abandoned its subordinate role and has gone on an adventurous quest of knowledge for its own sake. The result has been a relentless advance towards our technological civilization which is now threatening to get out of control. We now face an environmental crisis with its triple apocalypse of overpopulation, pollution of the environment, and the depletion of the Earth's natural resources.

But loss of control over our cultural creations is not the only sphere where human reason has gone astray. Intellectual activity has also led to religious and philosophical aberrations, in the form of language abstractions, resulting from the misuse of speech and writing. In the course of such linguistic acrobatics certain abstractions have become separated from their formerly material basis and have assumed a ghostly existence of their own. The result has been the emergence of metaphysical and religious phenomena and of abstract philosophy. These mental abominations have originated from natural and historical coincidences which can partly be reconstructed from early history. The religious beliefs and ideologies which are the result have played a regrettable part in human affairs. They have diverted human thought and action from man's principal objective – the preservation of the human species. The new assessment of natural selection allows us to recognize the situation and do something about it.

It is, however, questionable whether the decline of the human species towards extinction can be averted, even if the remedies indicated in this book are applied. What we can perhaps attempt is to bring back human life to a saner level, uncluttered by historical doubts and sophisticated notions, which have arisen from our wrongly deduced, paranoic evaluation of our status in nature and the universe. Such an effort requires the application of a large dose of common sense, aiming at the world-wide liberation of political and educational institutions from existing ideological and doctrinaire fetters. In spite of likely failure, the overall objective must still remain the preservation of our species.

Appendix

AN ANALYSIS OF THE TEN ARISTOTELIAN OPPOSITES
(from Aristotle, *Metaphysics*, i, ch. 5-8)

The ten opposites, according to Aristotle, are the 'odd' and 'even', 'limit' and 'unlimited', 'one' and 'many', 'right' and 'left', 'male' and 'female', 'rest' and 'motion', 'straight' and 'curved', 'light' and 'darkness', 'good' and 'evil', 'square' and 'oblong'.

First, a definition of the term 'opposite' or 'opposites'. According to Chambers' *Modern English Dictionary*, 'opposite' means standing in front, face to face, placed over against, that which is opposed or contrary.

1. *'odd'* and *'even':*

This obviously refers to alternative numbers in a continuous line. As an example, let us relate 'odds' and 'evens' to ten equal bottles placed in a line, and numbered one to ten. Here, the odd and even numbered bottles will alternate without any visible opposing principle. If we number the same bottles only by evens, 2, 4, 6 etc., or only by odds, 1, 3, 5 etc., the same effect will prevail. On the other hand if we like to oppose odd and even numbers by writing them opposite to each other on a piece of paper, we shall achieve a purely symbolic opposition without any further consequence. By placing odd against odd, and even against even, the same symbolic opposing effect is achieved. Consequently there is no inherent opposition in 'odds' and 'evens'.

2. *'limited'* and *'unlimited':*

The word limit, to be comprehensible, must be linked to a material entity that can be measured. 'Unlimited' merely means beyond the capacity of human measurement. 'Limit', therefore, refers to something sensorily perceptible, while 'unlimited' has only rhetorical significance. There is, therefore, no physical relationship between 'limited' and 'unlimited', and the two abstractions can be neither compared nor opposed.

3. *'One'* and *'many':*

These terms can make sense only if related to objects. Let us, for example, relate them to persons. First, we take 'one' person standing alone apart from a group of persons, which can either be called 'one' (i.e. one group), or alternatively 'many' (i.e. many persons).

If the one person joins the 'one' group, the group still remains 'one', but it also remains 'many'. This game could, of course, go on indefinitely. The conclusion is that whatever arrangement is made between 'one' and 'many', there is really nothing that is opposed between these two abstractions, even if the relationship is between 'one' person as contrasted with several ('many') persons.

4. *'Right'* and *'left':*

If referring to the feet of a person, the position of the right vis-à-vis the left foot is that they are placed side by side. On the other hand, if there are two people facing each other, each one's right foot will face the other's left foot. This could be construed as 'right' opposing 'left'. But if the same two people stand behind each other, each one's left foot will face the other's left foot, and each one's right foot the other's right foot; hence there will be no opposing principle involved.

5. *'Male'* and *'female':*

Assuming a woman has twins, one male and the other female: although it can be said that they differ in their sex, they cannot be considered as opposed to each other. As to differences of opinions between male and female, or to possible combatative situations, one can say that males are more often opposed by males than they are opposed by females. Male is just as much opposed to male as male is to female.

6. *'Rest'* and *'motion':*

A person sitting in a moving train is at rest and in motion at the same time. Is the person therefore opposing itself? On the other hand two bodies moving towards each other from opposite directions can clash. But if the motion of one body is in a direction away from the resting body (in the opposite direction), they can never meet. Therefore in a generalized way opposition between rest and motion is just as likely as it is unlikely.

7. *'Straight'* and *'curved'*:

Geometry sometimes points out that a straight line is part of a circle of infinite diameter (a curve). Hence, while metaphorically speaking the former is in the same relationship to the latter as a drop of water is to a river, no opposing principle is involved. One could as well say that a triangle is opposed to a rectangle, or any geometrical figure to any other geometrical figure.

8. *'Light'* and *'darkness'*:

Physically there exists no such thing as absolute darkness, and the term 'darkness' means only a relative state of light. But even if there were absolute darkness, no opposing principle can be construed for it in respect of light. A dark cow cannot be opposed to a light cow unless they are deliberately incited against each other. Equally, one could just as well incite a dark cow against another dark cow.

9. *'Good'* and *'evil'*:

As in the previous example, this 'opposition' can only be applied in a relative way. Genghis Khan has been called an evil man because he massacred his enemies wholesale; yet Field-Marshal Montgomery maintained that he did good by creating an orderly empire with relative well-being for the Mongols. To kill a person is morally evil, yet in warfare it is considered to be morally justifiable. As a general rule, therefore, the above moral evaluations depend on social conventions, which differ in different societies and at different times. There is, therefore, no opposing principle involved.

10. *'Square'* and *'oblong'*:

These so-called opposites fall into the same category as those in (7) above.

In conclusion the assessment made in the main text must be repeated, namely, that the play of Aristotelian opposites turns out to be a pure word game with abstractions having neither any scientific, philosophical, nor common-sense significance.

Notes

CHAPTER 1

1 Darwin, *The Origin of Species*, p. 4.
2 Ibid., p. 82.
3 Ibid., pp. 65-6.
4 Ibid., p. 66.
5 Rhodes, *The Evolution of Life*, p. 81.
6 Russell, *The World of the Soil*, p. 78.
7 Darwin, op. cit., p. 65
8 Kalmus, *Genetics*, p. 12.
9 Darwin, op. cit., p. 64.
10 Ibid., p. 76.
11 Rhodes, op. cit., pp. 182-3.
12 Darwin, op. cit., pp. 89-90.
13 Ibid., p. 230.
14 Ibid., p. 232.
15 Ibid., p. 386.
16 Ibid., p. 235.
17 Ibid., pp. 490-91.
18 Romer, *Man and the Vertebrates*, vol. I.
19 Howells, *Mankind in the Making*.
20 Rose, *The Chemistry of Life*.
21 Romer, op. cit., vol I.

CHAPTER 2

1 Childe, *Man Makes Himself*, pp. 23-4.
2 Maynard Smith, *The Theory of Evolution*, pp. 142-3.
3 Ibid.
4 Darwin, op. cit.
5 Rhodes, op. cit., pp. 107-9.
6 Maynard Smith, op. cit., p. 258.
7 Russell, op. cit., p. 258.
8 Romer, op. cit., vol. I, pp. 116-17.
9 Childe, op. cit., p. 33.
10 Solecky, quoted in J. R. Caldwell, *New Roads to Yesterday*, p. 103.
11 Barnett, *The Human Species*, p. 330.
12 La Barre, *The Human Animal*.
13 Maynard Smith, op. cit., p. 260.
14 Ibid., p. 261.
15 Ibid., p. 261.
16 Ibid., p. 261.
17 Kalmus, op. cit., p. 38.
18 Darwin, op. cit., p. 60.
19 Ibid., p. 70.
20 Cippola, *The Economic History of World Population*, pp. 76-86.

21 Ibid., p. 86.
22 Barnett, op. cit., p. 304.
23 Harrison, *Man the Peculiar Animal*, p. 294.
24 Cippola, op. cit., pp. 84-6.
25 Ibid.
26 Ibid.
27 Barnett, op. cit., p. 304.
28 *Observer*, London, 12.4.1970.

CHAPTER 3

1 Maynard Smith, op. cit.
2 *Observer*, London, 12.4.1970.
3 Darwin, op. cit., p. 85.
4 *Observer*, London, 12.4.1970.

CHAPTER 4

1 Von Bonin, *The Evolution of the Human Brain*.
2 Ibid.
3 Ibid.
4 Ibid.
5 Ibid.
6 Ibid.
7 Childe, op. cit.
8 Prof. E. Smith, *The Migrations of Early Culture*.
9 Childe, op. cit.
10 Von Bonin, op. cit.
11 Rhodes, op. cit.
12 Childe, op. cit.
13 Oakley, *Framework for Dating Fossil Man*.
14 Childe, op. cit.
15 Osborne, *Men of the Old Stone Age*.
16 Brace & Montague, *Man's Evolution*.
17 Maynard Smith, op. cit.
18 Cippola, op. cit.
19 Ibid.

CHAPTER 5

1 Childe, op. cit., p. 25.
2 Howells, op. cit.
3 Ibid.
4 Ibid.
5 Rhodes, op. cit.
6 Von Bonin, op. cit.
7 Osborne, op. cit.

CHAPTER 6

1 Lewis H. Morgan quoted in Evans Pritchard, *Sociology*.
2 Perry, *The Growth of Civilization*, p. 152.
3 Ibid., p. 54.
4 Medawar, *The Art of the Soluble*, p. 48.
5 Lévi-Strauss, *Structural Anthropology*, p. 3.

CHAPTER 7

1 Oakley, op. cit.
2 D. Bordes, *Upper Palaeolithic Cultures in Western Europe*, p. 139, in J. R. Caldwell, *New Roads to Yesterday*.
3 Professor Francis Bordes, *Mousterian Cultures in France*, pp. 77-8, in J. R. Caldwell, ibid.
4 Professor Francis Bordes in J. R. Caldwell, ibid.
5 Klima, *Middle Europe*, p. 203, in J. R. Caldwell, ibid.
6 Mellaart, *The Earliest Settlements in Western Asia*, p. 11.
7 Ibid., p. 25.
8 Okladinov, *Temperate Continental Asia*, p. 270, in Bridewood and Willey, *Courses Towards Urban Life*.
9 Griffin, quoted by Bridewood and Willey, ibid.
10 Emery, *Archaic Egypt*.
11 Elliot Smith, *Egyptian Mummies*, 1., pp. 113 and 120.
12 Man, *Royal Anthropological Institute*, March 1969.
13 Ibid., p. 26
14 Mellaart, op. cit.
15 C. Gorman, *Cultural Adaptations to the Topography of Plants and Animals of Mainland South-East Asia*.
16 Childe, op. cit., pp. 92-3.
17 Mellaart, op. cit., p. 4 – At the time of writing, new claims for earlier agricultural origins at the Spirit Cave in Thailand, made by Prof. W. G. Solheim, had not been confirmed.
18 In case of Thailand agricultural origins proving to be earlier, the sequence would have to be changed to Thailand–West-Asia, and Thailand–China ... etc. ... as before.
19 Gorman, op. cit., p. 29.
20 'The Beginning of Agriculture in the Far-East', *Antiquity*, September 1970, p. 282.

CHAPTER 8

1 G. Daniel, *The First Civilizations*, p. 13.
2 *The New Diffusionist*, on Jairazbhoy Lecture (Kraus) no. 7, 1972.
3 Childe, op. cit., p. 43.
4 Mellaart, op. cit., p. 176.
5 G. Daniel, op. cit., p. 176.
6 Ibid., p. 177.
7 G. Daniel, *The Idea of Prehistory*, pp. 90-91.
8 G. Daniel, *The First Civilization*, p. 175.
9 G. Daniel, *The Idea of Prehistory*, pp. 90-91.
10 G. Daniel, *The First Civilization*, p. 176.
11 Ibid., p. 81.

12 Ibid., p. 103.
13 Ibid., p. 81.
14 G. Daniel, *The Idea of Prehistory*, p. 90.
15 *The New Diffusionist*, no. 7, pp. 53-5.
16 Forde, *Habitat, Economy and Society*, p. 141.
17 Ibid., p. 76.
18 Ibid., p. 63.
19 Ibid., p. 430.
20 Mellaart, op. cit., p. 16.
21 Ibid., p. 26.
22 Ibid., p. 41.
23 Ibid., p. 84.
24 Childe, op. cit., p. 137.
25 G. Daniel, *The First Civilizations*, p. 67.
26 Childe, op. cit., p. 157.
27 The proposition that religion in general and the worship of gods in particular is of distinct historical origin and that it is not a natural and universal trait of human nature resulting from man's organic (psychological) evolution, contradicts many cherished assumptions of established anthropology. It is therefore bound to be strongly contested. On the other hand, the fact of a longstanding period of human evolutionary stagnation which has been operative for hundreds of thousands of years, hardly allows any other explanation. This does not mean that the Egyptian origin of divine kingship (as suggested by Smith and Perry) is conclusively proved, or that religion can not have originated elsewhere. All we can say is that the Egyptian version is the most likely working thesis for the origin of gods and religion, as far as available evidence goes. Claims for the existence of religious practices and the worship of gods or goddesses, like those advanced by Mellaart for Catal Hayuk in Turkey, are entirely unproven, and must be considered as pure conjecture, (see Mellaart, *Earliest Civilization of the Near East*, p. 77).
28 This contention is supported by the antiquity of the Egyptian calendar, the control of which was a prerogative of the god-king. Gordon Childe writes: 'Now, historical kings in Egypt as in Babylonia and elsewhere, were intimately connected with the regulation of the calendar. It is suggested that they owed their authority in part at least to the prehistoric genius who determined the solar year with the aid of Sirius. His discovery would have enabled him to predict the flood to the fellahin, and thus vindicate its magical power of controlling the seasons and the crops' (see *Man Makes Himself*, p. 137). This means that one of the first controllers of Egyptian irrigation was presumably the first god-king.
29 G. Daniel, *The First Civilizations*, p. 92.
30 D. J. Mulvaney, 'Prehistory of the Australian aborigine', *Scientific American*, San Francisco, March, 1966.
31 W. Solheim II refering to the continental shelf, 1972.
32 The arri can conceivably be identified as Perry's, *Children of the Sun*, being related to ruling families of Asiatic, (Indian) origin, which carried the system of divine kingship of originally Egyptian making and all it stands for to the islands of the Pacific.
33 Forde, op. cit., pp. 216-19.
34 Daniel, *The First Civilizations*, p. 174.
35 Ibid.

CHAPTER 9

1 Perry, op. cit.
2 Ibid.

CHAPTER 10

1 Lorenz, *On Aggression.*
2 Drever, *Dictionary of Psychology.*

CHAPTER 12

1 Williams, *Man and Monkey.*
2 Koestler, *The Ghost in the Machine,* p. 304.
3 Ardrey, *African Genesis.*
4 Ibid.
5 Perry, *The Growth of Civilization.*
6 Ibid.
7 *Daily Times,* Lagos, Nigeria.

CHAPTER 13

1 Maynard Smith, op. cit., p. 312.

CHAPTER 14

1 Kalmus, op. cit., p. 12.
2 Darwin, *The Variations of Animals and Plants under Domestication.*
3 Darwin, *The Origin of Species.*
4 Koestler, *The Act of Creation,* p. 490.
5 Ibid., pp. 491-2.
6 Carthy, *Animal Behaviour.*
7 Romer, op. cit., vol. I, pp. 58-9.

CHAPTER 15

1 Darwin, *The Origin of Species.*
2 *Time* Magazine, New York, 2.2.1970.
3 *Der Spiegel, Hamburg,* 1969, no. 34.
4 *Observer,* London, 1970.

CHAPTER 16

1 Koestler, *The Act of Creation,* p. 420.
2 Lockley, *Animal Navigation,* p. 61.
3 Ibid.

4 Ibid.
5 Von Bonin, op. cit.
6 Childe, op. cit., p. 29.
7 Tylor, *Anthropology*.
8 Ibid.
9 Ibid.
10 Childe, op. cit., pp. 195-?.
11 Perry, *The Primordial Ocean.*
12 Ibid.
13 Perry, *The Growth of Civilization*, pp. 476-9.
14 Ibid.

CHAPTER 17

1 Childe, op. cit., p. 54.
2 Howells, op. cit., p. 259.
3 Ucko, *Anthropomorphic Figurines of Predynastic Egypt and Neolithic Crete with comparative Material from the Prehistoric Near East and Mainland Greece.*
4 Perry, *The Primordial Ocean*, pp. 142-3.
5 Ibid.

CHAPTER 18

1 Koestler, p. 228.
2 Herodotus, *The Histories.*
3 Childe, op. cit., p. 209.
4 Herodotus, op. cit.
5 Aristotle, *Metaphysics*, 1987b, 18-25, in Santanilla, *The History of Scientific Thought.*
6 *Encyclopaedia Britannica*, 1961 vol.
7 Aristotle, op. cit., (i., ch. 5-8) in Santanilla, op. cit.
8 *Encyclopaedia Britannica*, 1961 vol.

Bibliography

ABBIE, A. A.: *The Original Australians,* Friederich Müller, London, 1969

'Criteria for the Comparison of Human Skulls', *Nature,* London, 1963

ALLEN, Grant: *Evolution of the Idea of God,* Watts, London, 1963

ARDREY, Robert: *African Genesis,* Fontana, London, 1967

ANATI: *Palestine before the Hebrews,* Jonathan Cape, London, 1963

BARNETT, Anthony: *The Human Species,* Collins, London, 1964

BARRE, Weston La: *The Human Animal,* Penguin Books, Harmondsworth.

BAUMGARTEL, E. J.: *Predynastic Egypt,* Cambridge University Press, 1965

The Cultures of Prehistoric Egypt, Oxford University Press, 1947

BIRNEY, R. B. and R. C. TEAVEN: *Instinct: a Composium,* V. van Nostrant Co., New Jersey, U.S.A., 1961

BONIN, Gerhard von: *The Evolution of the Human Brain,* University of Chicago Press, 1963

BONWICK, James: *The Last of the Tasmanians,* Sampson Low, London, 1870

BOTTOMORE, T. B. and M. RUBEL: *Karl Marx,* Penguin Books, Harmondsworth, 1961

BRACE, C. and Ashley MONTAGUE: *Man's Evolution,* Collier-Macmillan, London, 1968

BRIDEWOOD and WILLEY: *Courses Towards Urban Life,* Edinburgh University Press, 1962

BROWN, R. Radcliffe: *The Andamanese Islanders,* London, 1932

CALDWELL, J. R.: *New Roads to Yesterday,* Thames & Hudson, London, 1966

CAMERON, J.: *A Contribution to the Morphology of the Human Skull,* Royal Society of Canada, Ottawa, 1918

CARTHY, A. J.: *Animal Behaviour*

CARTER, C. O.: *Human Heredity,* Harmondsworth, Penguin Books, 1967

CHERRY, Professor T.: 'The Discovery of Agriculture', *Australian Society for the Advancement of Science,* Melbourne, 1921

The Origin of the Idea of the Supernatural, Melbourne, 1921

CHICAGO, Fildiana: *Cranial Capacities: A Study of Methods in Anthropology,* London, February 1947

CHILDE, E. Gordon: *Man Makes Himself,* Fontana Library, Collins, London, 1966

CIPPOLA, C. M.: *The Economic History of World Population,* Penguin Books, Harmondsworth, 1965

COE, M. D.: *The Maya*, Thames & Hudson, London, 1962
COON, C. S.: *The History of Man*, London, 1955
CYRIAX, Dr James: *The Slipped Disc*, Gower Press, London, 1970
DANIEL, Glyn: *The Megalithic Builders of Western Europe*, Penguin Books, Harmondsworth, 1958
The Idea of Prehistory, Penguin Books, Harmondsworth, 1964
The First Civilizations, Penguin Books, Harmondsworth, 1971
DARLINGTON, C. D.: *Genetics and Man*, Penguin Books, Harmondsworth, 1964
DARWIN, Charles: *Darwin on Humus and the Earth Worms*, Faber, London, 1945
The Origin of Species, World's Classics, Oxford University Press, London, 1958
The Descent of Man, John Murray, London, 1871
The Variations of Animals and Plants under Domestication, John Murray, London, 1868
DIXON, R. B.: *Contacts with America across the Southern Pacific*, Toronto, 1933
'The Long Voyages of the Polynesians', *Proceedings of the American Philosophical Society*, 1934
DREVER, James: *Dictionary of Psychology*, Penguin Books, Harmondsworth, 1968
EMERY, W. B.: *Archaic Egypt*, Penguin Books, Harmondsworth, 1967
EVANS, Pritchard, Professor E.: *Social Anthropology*, B.B.C. Lectures, Cohen and West, 1951
EYSENCK, H. J.: *Uses and Abuses of Psychology*, Penguin Books, Harmondsworth, 1963
FORDE, Daryl C.: *Habitat Economy and Society*, Methuen, London, 1968
FRANKFORT, Professor: *The Problem of Similarities in Ancient Near Eastern Religions*, Frazer Lecture 1950, Clarendon Press, Oxford, 1951
FRAZER, D.: *Early Art in the Pacific Basin*, Columbia University U.S.A., 1967
FRAZER, Sir J. G.: *The Golden Bough*, Macmillan, London, 1922
FREUD, Sigmund: *Totem and Tabu*, Routledge, London, 1950
GORMAN, Chester, F.: *Cultural Adaptations to the Topography of Plants and Animals of Mainland South-East Asia*, University of Hawaii, Honolulu, 1971
HAMBLY, W. D.: *Cranial Capacities: A Study in Methods*, Chicago Natural History Museum, 1947
HARDEN, P.: *The Phoenicians*, Thames & Hudson, London, 1962
HARRIS, M.: *The Rise of Anthropology*, Routledge & Kegan Paul, London, 1968
HARRISON, A. J.: *Man the Peculiar Animal*, Penguin Books, Harmondsworth
HENSCHEN, F.: *The Human Skull*, Thames & Hudson, London, 1966
HOMET, M. C.: *On the Trial of the Sun Gods*, Neville-Spearman, London, 1965

HOCART, W. M.: *Kingship*, Oxford, 1927

HERODOTUS: *The Histories*, Penguin Books, Harmondsworth, 1954

HOWELLS, William: *Mankind in the Making*, Penguin Books, Harmondsworth, 1967

HUXLEY, Julian: *Evolution – the Modern Synthesis*, Allen & Unwin, London, 1948

HEYERDAHL, Thor: *American Indians in the Pacific*, Allen & Unwin, London, 1952

JACKSON, J. W.: *Shells as Evidence of the Migration of Early Culture*, Longman Green & Co., London, 1917

KALMUS, H.: *Genetics*, Heinemann, London, 1964

KIANG KANG HU: *Resemblance of the Maya Civilization and that of the Chinese*, Ottawa, 1933

KOESTLER, Arthur: *The Ghost in the Machine*, Hutchinson, London, 1967

The Act of Creation, Hutchinson, London, 1964

KRAMER, S .W.: *History begins at Sumer*, Thames & Hudson, London, 1958

Enki and Ninhursag, Thames & Hudson, London, 1948

KRAUS, G.: *The New Diffusionist*, vols I and II, Gt. Gransden, 1971 and 1972

LEVI-STRAUSS, Claude: *Structural Anthropology*, Allen Lane the Penguin Press, London, 1968

LIGHTWALL, W. D.: 'The Origin of the Maya Civilization. Can China contribute to its Solution?' from *Transactions of the Royal Society of Canada*, Ottawa, 1933

LOCKLY, R. M.: *Animal Navigation*

LORENZ, Dr K.: *On Aggression*, Methuen, London, 1966

LOWIE, R. H.: *The History of Ethnological Theory*, Holt, N.Y., 1937

LURIA, Professor A. R.: *The Mind of a Mnemonist*, Jonathan Cape, London, 1969

MCBURNEY, C. B. M.: *The Stone Age of Northern Africa*, Penguin Books, Harmondsworth, 1960

MEDAWAR, P. B.: *The Art of the Soluble*, Penguin Books, Harmondsworth, 1969

MELLAART, James: *The Earliest Settlements in Western Asia*, Cambridge University Press, 1967

Earliest Civilizations of the Near East, Thames & Hudson, London, 1965

MILLS, C. Wright: *The Marxists*, Penguin Books, Harmondsworth, 1963

MONGIT, A. L.: *Archaeology in the U.S.S.R.*, Penguin Books, Harmondsworth, 1961

MURRAY, Margaret: *The Splendour that was Egypt*, Sedgwick & Jackson, London, 1949

OAKLEY, K. P.: *Framework for Dating Fossil Man*, Weidenfeld & Nicolson, London, 1964

OLIVER, R. and J. D. FAGE: *A Short History of Africa*, Penguin Books, Harmondsworth, 1965

OSBORNE, H. F.: *Men of the Old Stone Age*, Bell & Sons, London, 1926

OETTERING, B.: *Human Craniology*, Chirop. Inst. New York, 1957

PERRY, J. W.: *The Growth of Civilization*, Methuen, London, 1923

The Children of the Sun, Methuen, London, 1923

The Primordial Ocean, Methuen, London, 1935

The Origin of Magic and Religion, Methuen, London, 1923

RAGLAN, Lord: *How Came Civilization?*, Methuen, London, 1939

Some Aspects of Diffusion, Presidential Address, Royal Anthropological Institute, 1965

RHODES, F. T. H.: *The Evolution of Life*, Penguin Books, Harmondsworth, 1962

ROMER: *Man and the Vertebrates*, vol. I, Penguin Books, Harmondsworth, 1964

ROSE, Steven: *The Chemistry of Life*, Penguin Books, Harmondsworth, 1966

ROYAL ANTHROPOLOGICAL INSTITUTE: *Man*, (quarterly publication), London, March 1969

RUSSELL, Sir E. John: *The World of the Soil*, Penguin Books, Harmondsworth, 1966

SANTILIANA, O., de: *The History of Scientific Thought*, Mentor Books, Chicago, 1961

SCHUSTER, Carl: 'Joint Marks, a possible Index of Cultural Contact between America, Oceania and the Far East', *Royal Anthropological Institute*, Amsterdam, 1951

SMITH, Professor Elliot: *Human History*, Jonathan Cape, London, 1930

The Diffusion of Culture, Watts & Co., London, 1933

In the Beginning, Watts & Co., London, 1932

The Migrations of Early Culture, Manchester University Press, 1929

SMITH, John Maynard: *The Theory of Evolution*, Penguin Books, Harmondsworth, 1966

SOLHEIM, Professor W. G.: 'An Earlier Agricultural Revolution', in the *Scientific American*, April, 1972

STALIN, J. V.: *Collected Works*, Moscow.

THOMSON, J. E. S.: *The Rise and Fall of Maya Civilization*, Norman, Oklahoma, 1959

TAYLOR, A. J. P.: Introduction to the *Communist Manifesto*, Penguin Books, Harmondsworth, 1967

TYLOR, E. B.: *Anthropology*, John Murray, London, 1905

Primitive Culture, John Murray, London, 1903

WADDINGTON, Professor: *An Introduction to Modern Genetics*, Penguin Books, Harmondsworth, 1967

WILLIAMS, Leonard: *Man and Monkey*, Panther Science, London, 1969

UCKO, Peter: *Anthropomorphic Figurines of Predynastic Egypt and Neolithic Crete with comparative Material from the Prehistoric Near East and Mainland Greece*, Andrew Szmidla, London, 1968

Index

The New Diffusionist

A quarterly journal of comparative cultural anthropology linking the historical approach with the diffusionist school.

Published by The New Diffusionist Press, Editor: C. E. Joel

Designed to bring the diffusionist thesis up-to-date and to restore historical perspectives in all fields of cultural and social anthropology, *The New Diffusionist* subjects contemporary research to critical review and analyses through a continuing dialogue and some speculation the diffusionist model in the light of recent advances.

Contributions welcome supporting or constructively criticizing the editorial position: letters, articles and reports of recent investigations are invited.

Rigorous inter-disciplinary work unfettered by the narrow confines of specialization. International search for interesting material.

For further details please write to:
The Editor, The New Diffusionist, 39 West Street, Great Gransden. Sandy. Bedfordshire. England.
Published: January, April, July, October.
Price: 20p per copy. Subscription £1 or $3 for 6 copies (plus airmail surcharge).